Information Fusion and Data Science

Series editor
Henry Leung, University of Calgary, Calgary, Alberta, Canada

This book series provides a forum to systematically summarize recent developments, discoveries and progress on multi-sensor, multi-source/multi-level data and information fusion along with its connection to data-enabled science. Emphasis is also placed on fundamental theories, algorithms and real-world applications of massive data as well as information processing, analysis, fusion and knowledge generation. The aim of this book series is to provide the most up-to-date research results and tutorial materials on current topics in this growing field as well as to stimulate further research interest by transmitting the knowledge to the next generation of scientists and engineers in the corresponding fields. The target audiences are graduate students, academic scientists as well as researchers in industry and government, related to computational sciences and engineering, complex systems and artificial intelligence. Formats suitable for the series are contributed volumes, monographs and lecture notes.

More information about this series at http://www.springer.com/series/15462

Hao Jiang • Qimei Chen • Yuanyuan Zeng
Deshi Li

Mobile Data Mining and Applications

 Springer

Hao Jiang
School of Electronic Information
Wuhan University
Wuhan, China

Qimei Chen
School of Electronic Information
Wuhan University
Wuhan, China

Yuanyuan Zeng
School of Electronic Information
Wuhan University
Wuhan, China

Deshi Li
School of Electronic Information
Wuhan University
Wuhan, Hubei, China

ISSN 2510-1528   ISSN 2510-1536 (electronic)
Information Fusion and Data Science
ISBN 978-3-030-16502-4   ISBN 978-3-030-16503-1   (eBook)
https://doi.org/10.1007/978-3-030-16503-1

This Springer imprint is published by the registered company Springer Nature Switzerland AG.
The registered company address is: Gewerbestrasse 11, 6330 Cham, Switzerland

Preface

With the development of computer science technology and communication technology in recent years, people's ability of collecting and producing data using information technology has increased dramatically. Various types of applications built on this foundation have also emerged. At the same time, with the popularity and development of mobile device technology, mobile data is exploding. Correspondingly, people are also facing the technical challenge of massive data and excessive information. Mobile data mining aims to extract the hidden, uninformed, and potentially valuable information from these large amounts of unknown data. The application fields of mobile data mining are all areas that require knowledge management and decision-making. These areas are faced with complex environment and application requirements. Different types of applications need different information to assist. For example, in the field of mobile communication, mobile data mining can be used to understand the user's behavior patterns and habit preferences, so as to understand the user's social association and lifestyle, and serve the featured application scenarios. From the perspective of technological development, mobile data mining will be an ever-developing and progressive science and technology in the field of computer science and communications. It has strong vitality and broad application prospects and also plays an important role in social life and production.

The content of this book is organized as follows: Chap. 1 is an overview of the evolution of mobile data mining and application technology. It is intended to give readers a basic understanding of the technological development process in mobile data mining and application areas in recent years. Chapter 2, Mobile Data Processing and Feature Discovery, illustrates the basic methods and feature discovery methods of mobile data processing from two aspects: Mobile Data Sensing and Feature Discovery of Mobile Data Users and Networks. Chapter 3, Mobile Data Application in Mobile Communications, illustrates the key technologies of mobile data applications from the perspective of mobile communications. Chapter 4 is the Mobile Data Application in Mobile Network. The mobile data mining and application technologies in mobile networks are described from three aspects: Mobile Data Application in D2D Network, Mobile Data Application in

Green Communication Network, and Mobile Data Application in Sensing Network. Chapter 5 is Mobile Data Application for Smart City. Chapter 6 summarizes current mobile data mining and application technologies.

Thanks to the teachers and students of Advanced Network and Intelligent System (ANTIS) Research Team of Electronic Information School, Wuhan University, for their support. Thanks to the postgraduates Shujie Zhou, Cong Zheng, Xiaoyue Zhao, and other students for their hard work in the writing and typesetting. Thanks to Yuanyuan Zeng, Qimei Chen, and other teachers in the team for writing the most part of the book and providing a lot of valuable advice. This book is intended to be shared and communicated with more researchers and students. The publication of this book has received strong support from Springer, and I would like to express my sincere gratitude. In the process of writing this book, I also want to express my heartfelt thanks to the teachers and students in communication center of the Geospatial Information Technology Collaborative Innovation Center and the Electronic Information School of Wuhan University.

Due to the limited time and ability, there are inevitable shortcomings and omissions in the book. We welcome experts and readers to criticize and correct.

Wuhan, China   Hao Jiang

Acknowledgments

This work was supported in part by the National Natural Science Foundation of China (NSFC), Grant No. 61702387 and No. 61801333; the National Key Research and Development Program, No. 2017YFC0503801; 2018 Shanghai Artificial Intelligence Innovation Development Special Support Project (2018-RGZN-01013); the Fundamental Research Funds for the Central Universities, No. 2042019kf1001; and 2018 Ministry of Industry and Information Technology Big Data Industry Development Pilot Demonstration Project.

Contents

Chapter 1
Introduction

Abstract With the development of wireless communication, mobile data is becoming more widespread. At the same time, mobile data analysis and mining come into being. Also, there are many applications based on mobile data analysis and data mining technology. This chapter is the introduction part of the whole book, which mainly describes mobile data and purpose, methods, applications of mobile data analysis, and mining.

Keywords Mobile terminals · Mobile data analysis · Data mining

With the popularity of mobile terminals such as smartphones, tablets, and notebook computers, mobile traffic has grown exponentially in recent years. The rapid development of the mobile Internet and the popularity of mobile terminals will lead to an eightfold increase in mobile data traffic in 2020, and bandwidth-sensitive applications such as video will account for more than 70% of all data requests. According to Cisco's mobile network outlook report, global mobile data traffic will grow to 292 EB in 2019, and smart traffic accounts for up to 97%.

Fast-growing wireless location technology helps people to access mobile data easily. Mobile data can be used not only for positioning, tracking, but also for representing spatial information [1]. With the popularity of various handheld devices such as GPS, mobile phone, and so on, people can record their mobile data at any time and arrange mobile data in chronological order. On analyzing the mobile data we can get a lot of potential information. Mobile data collection channels are diverse, and most of the mobile data mainly record users' time and space information which are important components of people's daily activities. Therefore, mobile data has many applications. Some applications of mobile are as follows: First, we can use mobile data to mine users' mobile behaviors and find users' living habits. Second, predict users' possible behaviors based on users' mobile behaviors. Third, on analyzing the mobile data we can get many advanced information, such as users' interests, hobbies, and social relationship between users.

There are several fast-moving revolutions in the technological landscape in the past few years, particularly related to the emergence of powerful, always-

© Springer Nature Switzerland AG 2019   1
H. Jiang et al., *Mobile Data Mining and Applications*, Information Fusion and Data Science, https://doi.org/10.1007/978-3-030-16503-1_1

connected, and extremely popular portable devices. Smartphones and tablets let us receive information through multiple channels while generating massive amounts of information about us. Data collected from the sensors embedded in smartphones especially GPS receivers provide an incredible wealth of information that service providers and applications can collect, store, and analyze in real time. Computing is truly personal now, not only because we access information through mobile devices, but also because the information itself is usually highly personalized and relevant to our location and context. Typical examples are location-based services, such as Foursquare, which provide suggestions about restaurants and shops close to the area where users have checked in, and consider their previous mobility history. Other examples include search engines that are increasingly context and location-aware. Moreover, users generate information themselves using mobile devices. For example, in June 2013, Facebook had, on average, 819 million monthly active mobile users. In other words, we should be talking not about the big data revolution but at least as far as consumer applications are concerned about the "big mobile data" phenomenon [2]. However, another trend will be even more central in the years to come: big mobile data are increasingly used not just for analyzing the past or understanding the present, but also for predicting the future. A new paradigm is slowly emerging that we might define as anticipatory mobile computing [3], examples include companies that are developing innovative mining applications for real-time marketing or for supporting strategic decisions in retailing, such as the Telefonica Smart Steps project.

At present, analysis of mobile data mainly uses data mining technology. Therefore, mobile data analysis and mining come into being. At present, intelligent electronic equipment appears in the daily life of everyone. With the development of electronic equipment, the number and means of information received by people have steadily increased over the past few years. These information which can be gotten easily can satisfy our need but also can trouble our life because of falseness of information. In order to avoid the presence of adverse reactions, data analysis and data mining are widely applied in communication field. Data mining services play an important role in the field of communication industry. Data mining is also called knowledge discovery in several database including mobile databases [4]. According to related data statistics, data mining technology is a cross-disciplinary technology, which covers statistics, database technology, visualization, information science, machine learning, and so on. Therefore, people can achieve the retrieval of information and extract valid information in order to provide technical support for mobile communication. The mobile communication industry has made great progress in the process of development of information. Various application systems are widely applied, such as billing system and integrated system. These systems can save full historical data but also can make data redundant which cause users not to extract useful information. Moreover, massive data bring troubles in operating effectively for computer equipment. Data mining technology helps users satisfy the need and decrease cost as well as improve efficiency.

Many methods are applied in mobile data mining. Some key technologies are listed next. First, build database. People build a platform that can share data of mobile communication services, use appropriate algorithms to mine data information, obtain useful information, and provide scientific and technical support for the development of operators. Second, apply decision tree count. The technology generates a corresponding model for the marketing of goods with different characteristics in order to differentiate customer segments effectively and use different strategies for different groups. Third, association rules, which are used to alarm correlation analysis and search for the point of failure and the cause fast. Also it can make maintenance more convenient and improve work efficiency.

There are many key technology in mobile data mining and analysis including feature discovery technology, perceptual data acquisition, and communication technology such as big data acquisition related technology of Internet of things (IoT) and D2D technology as well as green communication technology, application of mobile data smart city which reveals the characteristics, patterns, and regulation of the city by obtaining city-related data.

Therefore, there are many applications based on mobile data analysis and data mining. Mobile data stream mining is particularly significant for applications that need real-time analysis of continuous data streams such as mobile crowd sensing, mobile activity recognition, intelligent transportation systems, mobile health care, and so on [5]. Mobile big data can be used to improve transportation in developing countries, devise strategies for epidemic containment [6], or study social response during major disasters such as earthquakes. Rather interestingly, given the fact that companies and other governmental and nongovernmental organizations own most of these data, a successful way to make them available for researchers and the public in general has been through challenges (such as the Orange Data for Development challenge) or via hackathon events, which usually attract vast number of programmers and researchers. Other applications include analyzing big data to understand the behavior and emotional states of groups of individuals [7], for example, to tackle mental health problems such as depression by devising effective behavior intervention strategies at group or population levels [8]. Another potential application for large-scale mobile data mining is building analytical and predictive applications for law and order enforcement. You would have already read titles in newspapers about the development of minority report-like systems [9]. Such headlines probably aren't true and shouldn't be considered worrying. Rather, most of these studies involve trying to identify crime hotspots in cities, analyze these hotspots' characteristics (for example, via census information), and identify any emerging geographic patterns. Besides, data mining technologies also are widely applied in mobile communication. Business forecasting identifies the key influencing factors of business development by analyzing users' historical information, including users' personal data and usage situation. Through reasonable prediction of relevant factors, the possible situation of future business development is determined so as to provide decision-making basis for the next business development, formulation of strategy, and so on. According to the actual situation, the prediction model and algorithm are improved. The main algorithms used are time series model and neural network prediction. Data mining can also predict and control user loyalty. "User loyalty" refers to the possibility of users changing operators. Through the analysis of off-line identification, user information data, account and payment data, and other data, we can measure user loyalty and take relevant measures to maximize user retention. By subdividing the user groups by clustering algorithm, it provides a basis for one-to-one marketing and the development of related new products. Data mining technology can also be applied to network resource management, key customer feature recognition, and other fields, through the analysis of relevant data, to provide decision-making for operators.

The rest of this book is structured as follows: In Chap. 2, we introduce mobile data processing and feature discovery. In Chap. 3, we introduce mobile data application in wireless communications. In Chap. 4, we introduce mobile data applications in mobile network. In Chap. 5, we introduce mobile data application in smart city. In Chap. 6, we give concluding remarks and future directions.

References

1. Lin, M., Hsu, W.: Mining GPS data for mobility patterns: a survey. Pervasive Mob. Comput. **12**(3), 1–16 (2014)
2. Musolesi, M.: Big Mobile data mining: good or evil? IEEE Internet Comput. **18**(1), 78–81 (2014)
3. Pejovic, V., Musolesi, M.: Anticipatory mobile computing: a survey of the state of the art and research challenges. ACM Computing Surveys (CSUR) Surveys Homepage table of contents archive (2015). https://doi.org/10.1145/2693843
4. Ghotkar, S.A., Shelke, R.R.: Implementation of effective mobile data mining through algorithm. Int. J. Adv. Res. Comput. Sci. **4**(2), 214–217 (2013)
5. Krishnaswamy, S., Gama, J., Gaber, M.M.: Mobile data stream mining: from algorithms to applications. In: 2012 IEEE 13th International Conference on Mobile Data Management (MDM), pp. 360–363 (2012)
6. Lima, A., De Domenico, M., Pejovic, V., Musolesi, M.: Exploiting cellular data for disease containment and information campaign strategies in country-wide epidemics. eprint arXiv:1306.4534 (2013)
7. Lazer, D., et al.: Computational social science. Science **323**(5915), 721–723 (2009)
8. Lathia, N., et al.: Smartphones for large-scale behavior change intervention. IEEE Pervasive Comput. **12**, 66–73 (2013)
9. Sgort, M.B., et al.: A statistical model of criminal behavior. Math. Models Methods Appl. Sci. **18**(1), 1249–1267 (2008)

Chapter 2
Mobile Data Processing and Feature Discovery

Abstract This chapter mainly describes mobile data processing and feature extraction methods. In the first section, we introduce the method of collecting and acquiring mobile sensing data. We mainly describe the human-centric crowd sensing method. There are two methods of collecting data called participatory data sensing and opportunistic data sensing introduced in this section. In the following sections, both the user feature discovery method and the feature extraction method of user group behavior discovery and context awareness are introduced.

Keywords Mobile data processing · Feature extraction · Sensing data · Human-centric · Participatory · Opportunistic · Behavior discovery · Context awareness

2.1 Mobile Data Sensing

In the last 10 years, the hotspots of data sensing research have been gradually advanced from static data sensing networks deployed in specific monitoring areas to perceptions of people's surroundings. Network nodes have also advanced from static sensor nodes to mobile nodes such as mobile sensing of vehicles, people, etc. Mobile sensing is designed to take advantage of the mobile capabilities of vehicles, airplanes, ships, people, etc., combined with ubiquitous sensing technology to provide low-cost, more flexible perceptions data collection via short-distance wireless communications for monitoring objects away from network infrastructure. In recent years, with the maturity of sensing technology and application environment, mobile sensing is mainly dedicated to providing people with effective monitoring of the surrounding urban environment, thus providing timely, effective, and fine-grained monitoring such as air pollution, noise pollution, and traffic congestion to improve people's quality of life and protect the environment which provides a "people-centric" perception relying on human mobility. People obtain data related to the city through information acquired in the surrounding environment, such as mobile terminals and vehicles. They also provide data integration, analyze and mine through data processing technology, acquire knowledge, and realize air pollution monitoring, water quality monitoring, noise monitoring, traffic congestion, and road driving planning effectively in urban areas.

Mobile data sensing provides ubiquitous sensing services for urban-scale applications through various connected smart terminal devices or network devices to achieve a close relationship between people and things. Mobile sensing is closely related to human mobility. Understanding "human mobility" is critical to mobile-aware web applications and related mechanism design. In the human-centric mobile date sensing network architecture, people, as managers of various types of intelligent sensing devices, largely determine the degree of participation and perceived effectiveness of perceived activities. According to people's subjective awareness and participation in perceived activities, mobile data sensing can be divided into participatory data sensing and opportunistic data sensing.

2.1.1 Participatory Data Sensing

2.1.1.1 Challenges for Participatory Sensing

Participatory sensing [1, 2] is a novel and promising sensing paradigm with the fast development of mobile terminals. Participatory sensing paradigm enables people with smart devices as the social sensors to be part of the sensing campaign to collect data from ambient environment. The sensing data then can be shared and analyzed to reveal a certain pattern of the city. Participatory sensing is well suited for applications such as air quality monitoring, noise monitoring, transportation monitoring, etc. The objective of participatory sensing is to supervise the process and results of social activities.

2.1.1.2 Behavior Modeling in Participatory Sensing

The behavior modeling includes the modeling of tempo-spatial related behavior and behavior tie.

We represent the mobile trajectory of people held with smartphones or other smart devices in participatory sensing using the association matrix as illustrated in Fig. 2.1. In the matrix, the row corresponds to a location grid. The location grid $g_0, g_2, \ldots, g_l$ is originated from the place division of the interested area for sensing task. The column corresponds to the time period at the location grid for typical time duration such as a day or a week. In people's daily life, behavior varies within different time period. The value of column is based on this. For example, time 24:00–6:00 is the rare moving time duration for most of the people in daily life. The other time durations like 6:00–8:00 and 8:00–10:00 show more activity of moving trajectory. The matrix elements represent the percentage of time that people stay inside the grid.

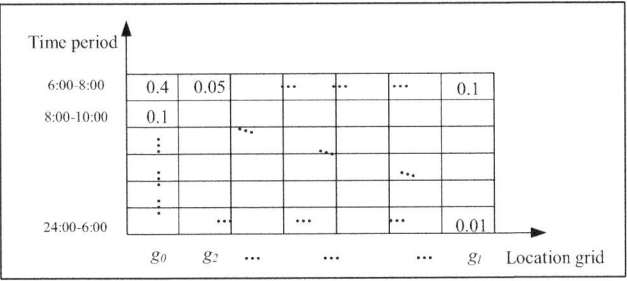

Fig. 2.1 The association matrix based on mobile trajectory

We utilize the method in [3] to capture the Eigen-behavior vector for a given participant. For participant x with association matrix A, the singular value decomposition (SVD)[4] is applied as in Eq. (2.1).

$$A = U \cdot \Sigma \cdot V^T \tag{2.1}$$

In which, a set of Eigen-behavior vectors representing the leading trends within the typical time period can be obtained from the column of U, i.e., the rows of matrix V^T. Σ is the diagonal matrix with corresponding singular values, $\sigma_1, \sigma_2, \ldots, \sigma_{\text{rank}(A)}$. The weight of each Eigen-behavior vectors can be achieved as in Eq. (2.2).

$$\omega_i = \frac{\sigma_i^2}{\sum_{j=1}^{\text{rank}(A)} \sigma_j^2} \tag{2.2}$$

The tempo-spatial similar behavior of two participants can be achieved through the cosine similarity of Eigen-behavior vectors. For users x and y with association matrix A_x and A_y, the similarity is calculated as shown in Eq. (2.3).

$$\text{sim}(x, y) = \text{sim}(U_x, U_y) = \sum_{i=1}^{\text{rank}(U_x)} \sum_{j=1}^{\text{rank}(U_y)} \omega_{xi} \omega_{yi} |U_{xi} \cdot U_{yi}| \tag{2.3}$$

The behavior of people in daily life is constrained by the social activities and affected by social relations. The user behavior tends to be with homophile, which explains the reason for building ties among similar users. There are strong tie and weak tie in social life that affect the substantial social participatory sensing campaign, i.e., the tie affects participation. We use information theory based methods to measure the correlation of participation to their probability of happening. The normalized mutual information (NMI) represents the normalized behavior correlation of two users in participatory sensing, i.e., behavior tie in participatory sensing. Based on this, the mutual information [5] metric measures the dependence of two users in terms of their behavior in participatory sensing. Given two random participating events X and Y with marginal probability mass function $p(x)$ and $p(y)$ representing the probability of the participation for user x and user y in a specific participatory sensing campaign, and the joint probability mass function of $p(x, y)$ representing the joint probability of the participation for user x and y, their mutual information $I(X; Y)$ is defined as the relative entropy between the joint distribution and their product distribution as in Eq. (2.4).

$$I(X; Y) = \sum_{x \in X, y \in Y} \log \frac{p(x, y)}{p(x)p(y)} \tag{2.4}$$

The NMI can be defined as in Eq. (2.5) where $H(X)$ and $H(Y)$ are the entropy of x and y. NMI value (i.e., behavior tie as we defined) is between 0 and 1, while value 1 implies X and Y with the same behavior in participatory sensing.

$$\text{NMI}(X; Y) = \frac{I(X; Y)}{\sqrt{H(x)H(y)}} \tag{2.5}$$

According to the historical record of participants for a certain long-time duration T, the user x moves from grid g_0 to g_1 with t_x as time fraction of participation. The user y moves from grid g_0 to g_1 with t_y as time fraction of participation. The entropy $H(X)$ and $H(Y)$ is shown in Eq. (2.6).

$$H(X) = -\sum_{g_0}^{g_i} \frac{t_x}{T} \log \frac{t_x}{T}$$
$$H(Y) = -\sum_{g_0}^{g_i} \frac{t_y}{T} \log \frac{t_x}{T} \tag{2.6}$$

Based on it, we have Eqs. (2.7) and (2.8).

$$I(X; Y) = \sum_{g_0}^{g_i} \frac{t_{x,y}}{T} \log \frac{\frac{t_{x,y}}{T}}{\frac{t_x}{T} \frac{t_y}{T}} \tag{2.7}$$

$$\text{NMI}(X; Y) = \frac{\sum_{g_0}^{g_i} \frac{t_{x,y}}{T} \log \frac{\frac{t_{x,y}}{T}}{\frac{t_x}{T} \frac{t_y}{T}}}{\sqrt{\sum_{g_0}^{g_i} \frac{t_x}{T} \log \frac{t_x}{T} \sum_{g_0}^{g_i} \frac{t_y}{T} \log \frac{t_x}{T}}} \tag{2.8}$$

2.1.1.3 SBR Scheme

We assume that all users with handheld smart devices can get to know sensing tasks from a sensing task distribution platform of service provider. The users who join the platform and agree to be recruited can become a participant. The recruitment is executed during users' daily life without intervening their daily activities. Next introduce recruitment recommendation and recruitment programming as well as self-adaptive strategy.

Data quality is a key factor to decide the recruitment in participatory sensing. From the history sensing data, some metrics such as data accuracy, redundancy, relevance, completeness, and timeliness can be considered to measure the data quality contributed, which can assist to build the participants' reputation. A final participant data quality rating score can be rated by the service provider and then be computed based on the weighted average metric value as shown in Eq. (2.9). In which, the metrics can be decided and chosen by service provider according to sensing requirements.

$$R = \alpha_1 \cdot \text{accuracy} + \alpha_2 \cdot \text{timeliness} + \alpha_3 \cdot \text{relevance} + \alpha_4 \cdot \text{completeness} +$$

$$\alpha_5 \cdot \text{timeliness}$$

$$(2.9)$$

Considering data quality is highly related with user behavior in sensing activities along with their daily life, we use recommendation in data quality prediction. Based on the behavior tie for users in participatory sensing, we can predict the unknown user behavior of participation by considering the user interactions. The data quality based on recruitment recommendation becomes a collaborative filtering problem. The behavior tie implies the correlation of user behavior in participatory sensing with the value between 0 and 1. Then we choose the users with correlated behavior to make recommendation for a specific sensing campaign i, i.e., with $\text{NMI}(X_i; Y_i) \geq \text{NMI}_{th}$. For user x who satisfied the above requirement, the recommendation with weighted policy is shown as in Eq. (2.10). In which, we get the recommended data quality rating score as PR_y, i with participation candidate y toward a specific sensing campaign i.

$$PR_{y,i} = \overline{R_y} + \frac{\sum \text{NMI}(X_i, Y_i) \cdot (R_{x,i}, \overline{R_x})}{\sum |\text{NMI}(X_i, Y_i)|} \qquad (2.10)$$

The recruitment metrics that we consider in the scheme include tempo-spatial behavior characteristics of participants, data quality that can be achieved through the recruitment strategy, and budget users who can afford. Based on this, the recruitment problem can be modeled as a linear programming optimization problem for recruitment objective. For a specific participatory sensing task $i \in I$ with the participants set $P^* = \{p_1, p_2, \ldots, p_n\}$, the recruitment programming can be formulated as shown in Eqs. (2.11)–(2.14).

$$\max \sum_{k=1}^{n} \text{sim}(p, A_{\text{sid},i}) \tag{2.11}$$

$$\text{sim}(p_{ki}, A_{\text{sid},i}) \geq \text{sim}_{th} \quad \forall p_k \in P^* \tag{2.12}$$

$$PR_{p_k,i} \geq R_{th} \quad \forall p_k \in P^* \tag{2.13}$$

$$\sum_{k=1}^{n} \text{budget}(p_k) \leq \text{bd}_{th} \tag{2.14}$$

In Eq. (2.11), $A_{\text{std},i}$ is the typical tempo-spatial association matrix that can be given by the upper-layer application or through a long-time observation for sensing activities with the required coverage of grids and time fractions. $\text{sim}(p_i, A_{\text{std},i})$ is the similarity for partition candidate pi with the required tempo-spatial coverage. We have $\text{sim}(p_i, A_{\text{std},i}) = \text{sim}(A_{pi}, A_{\text{std},i})$, where the similarity value is between 0 and 1. In which, 0 implies no similarity with the required tempo-spatial characteristics. 1 implies the exact required tempo-spatial characteristics. The objective of recruitment programming is to maximize the total tempo-spatial behavior similarity for participants, while it subjects to Eqs. (2.12)–(2.14). In Eq. (2.12), sim_{th} is the similarity threshold required for sensing task, which provides the required tempo-spatial behavior similarity for participants. In Eq. (2.13), R_{th} is the threshold for data quality rating required for sensing task i. It provides the constraint of data quality. In Eq. (2.14), bd_{th} is the budget threshold that users can afford. It provides the budget constraint in sensing task, battery energy, etc. The above threshold value is given and adapted by upper-layer applications.

Sensing data is the main concern for participatory sensing campaign. Data quality can reflect the effectiveness of recruitment. However, the recruitment performance with participants will change with varied people's behavior in daily life and activities. We utilize self-adaptive recruitment strategy to update the participants according to the dynamic scenarios. The self-adaptive scheme is to check the data quality periodically as interval parameter T_d. If the data quality for current participant $p_k \in P^*$ cannot satisfy the requirement for a specific check duration, i.e., $R_{pk} < R_{th}$, the unqualified participant will be removed. If the overall data quality for current participant set P^* cannot satisfy the requirement for a specific check duration, i.e., $R_{P^*} < R_{th}$, a new round of recruitment is involved to recruit new participants to provide the efficient data quality.

2.1.1.4   Evaluation

We implemented the proposed scheme by programming in Java to make performance evaluations. Our evaluation is based on real trajectory sets of people's daily life by using GeoLife dataset [6], which was a project of Microsoft Research Asia with 182 users over 5 years from Apr 2007 to Aug 2012. Through data analysis, we divide the trajectory area of GeoLife into 5874 grids according to the latitude

Fig. 2.2 The tensor illustration of GeoLife user data

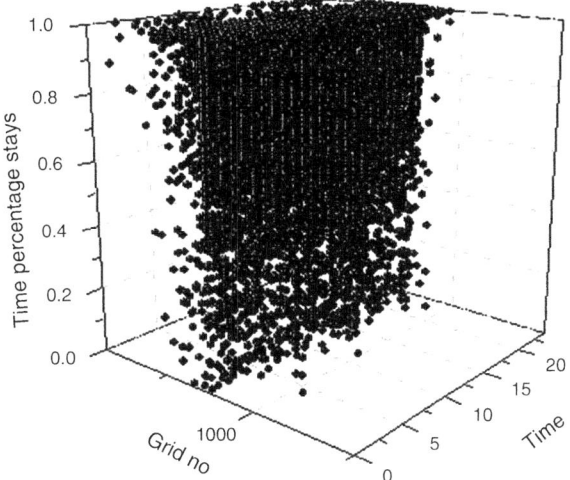

and longitude range of the trajectory. On the other hand, we divide the trajectory data into each hour according to the time within a day. Based on it, we extract the association matrix data for each user with grid number, hour number, and stay duration within the specific grid and hour. The data of 182 users in GeoLife dataset forms a tensor that can be illustrated as shown in Fig. 2.2.

The following section evaluates the scheme according to stability, low-cost, tempo-spatial correlation, and self-adaptiveness. As [7], our scheme is compared with three recruitment schemes: random, naive, and greedy. Random scheme selects participants randomly. Naive scheme selects participants who cover the grids and time hours without considering what existing selected participants covered. Greedy scheme selects participants who cover the most grids and time hours by considering what existing selected participants covered. We generate the target scenarios of participatory sensing campaign by choosing the random grids, random hours, and random stay duration within the most people's activity area and time period. Our evaluation results are based on the average of five groups of different random scenarios.

For a new participatory sensing campaign, the participant recruitment is based on the history trajectory. We use the GeoLife data before 2009 as the history trajectory. In a new sensing task, the data quality rating encounters the cold start. We assume that all participant candidates have the same initial data quality score that is qualified. We also assume that we can afford the budget during the evaluation. Figure 2.3 shows the number of recruited participants with our scheme SBR when compared with the other three schemes. It implies that SBR is with less cost, i.e., needs less to pay for the participants within the sensing requirements. From the result, we can see that SBR recruits the least number of participants for the target sensing scenario. Naive scheme recruits the most number of participants, because it only selects participants with the valid tempo-spatial coverage for the task scenario

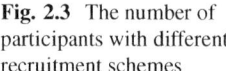

Fig. 2.3 The number of participants with different recruitment schemes

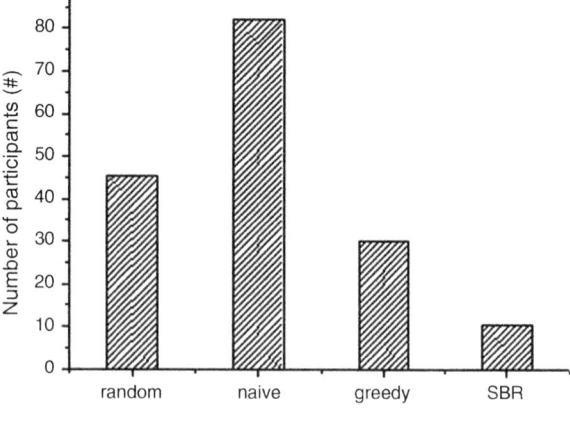

Fig. 2.4 The average stay duration with different recruitment schemes

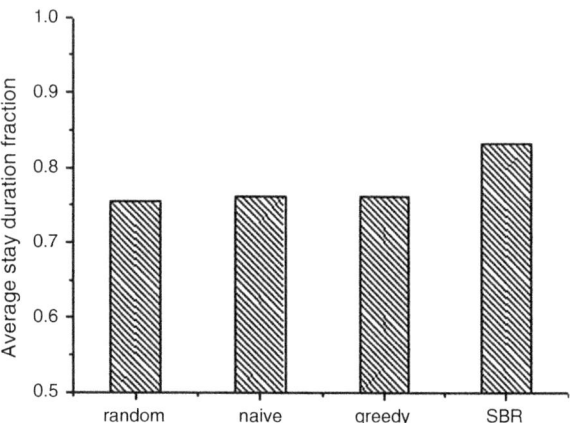

without considering the existing participants. Greedy scheme reduces the number of participants by considering the existing participants' coverage. Random scheme still recruits considerable participants by arbitrary selection. Figure 2.4 shows the average stay duration time fraction within the task grids of the recruited participants (named average stay duration fraction) with different recruitment schemes. The average stay duration implied the stability of participants for the sensing task, i.e., the longer the average stay duration, it means the participant stay longer in the expected grids within the expected time for the sensing task. It shows that people provide sensing data with more stable sensing habits, which helps to improve the data quality. The results show that SBR achieves the best average stay duration. It implies that SBR recruits participants with a relative good stability of tempo-spatial correlation for the sensing task. The other three schemes are with less average stay duration, and they have slight differences among one another.

We then made evaluations with different similarity threshold to see the effect of tempo-spatial correlation within different threshold selection. Figure 2.5 shows the participant number when we choose different similarity threshold sim_{th}. The results

Fig. 2.5 The number of participants with different similarity threshold

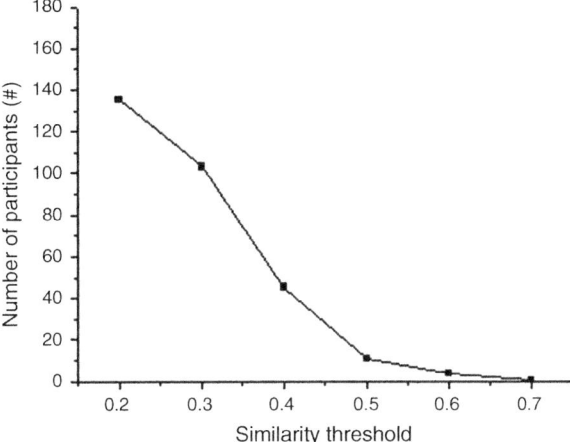

Fig. 2.6 The average stay duration with different similarity threshold

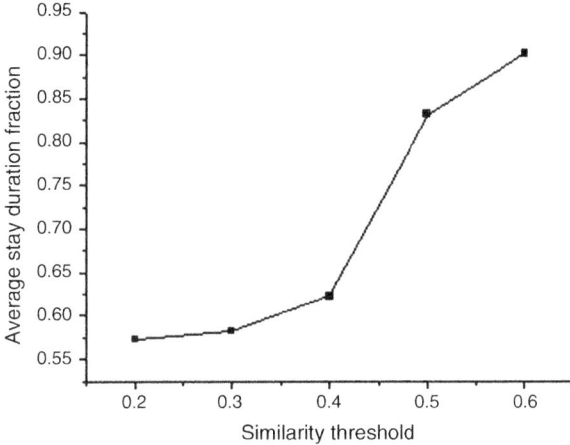

show that the participant number decreases as the similarity threshold increases. But the unreasonable similarity threshold incurs too many or too few participants. Too many participants increase the cost and incur the possible tempo-spatial redundancy, while too few participants cannot satisfy the tempo-spatial coverage. Figure 2.6 shows the average stay duration when choosing different similarity threshold sim_{th}. The results show that the average stay duration increases as the similarity threshold increases. As we set threshold $sim_{th} \geq 0.5$, the average stay duration gradually arrives at a considerable value than $sim_{th} < 0.5$. It implies that the tempo-spatial related sensing habits of people tend to be more valuable for sensing tasks when choosing $sim_{th} \geq 0.5$.

Based on the last subsection, we consider the campaign within a sustained scenario. We use the GeoLife data from the 1st to 10th month in 2009 as the assumed current on-time updated trajectory (considering the sparse trajectory after Oct. 2009 that may incur poor coverage, we choose data of the 1st to 10th month in 2009).

The update interval is taken as 1 month. The data quality is recommended after a random rating during the participatory sensing. In realistic situation, the data quality can be rated according to some combined metrics such as data accuracy, redundancy, and timeliness. Figure 2.7 shows the self-adaptive participant number that changes with the update interval, i.e., months. At the end of each update interval, a data quality based check is held to see if current data quality is qualified and needs to trigger the next new recruitment. The results show that SBR can self-adapt the strategy to recruit suited number of participants who satisfy the current application scenarios. Figure 2.8 shows the self-adaptive average stay duration, which can stay at a certain good level through self-adaptiveness. From the results, we can see that the recruitment provides self-adaptiveness with a relative stable sensing performance as the participant trajectory changes. It shows that SBR can adjust the recruitment strategy and provide a good sensing capability with participants with varied network scenarios.

Fig. 2.7 The self-adaptive participant number

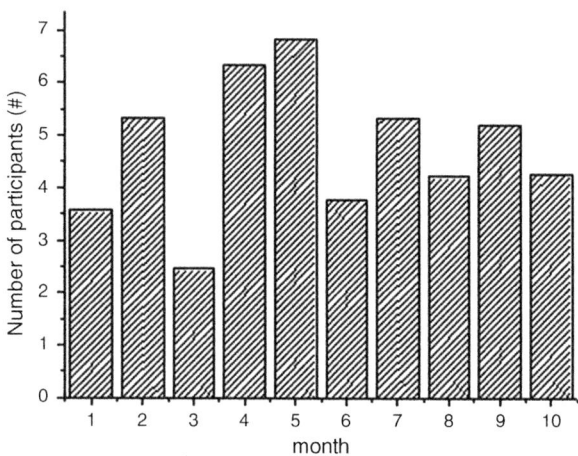

Fig. 2.8 The self-adaptive average stay duration

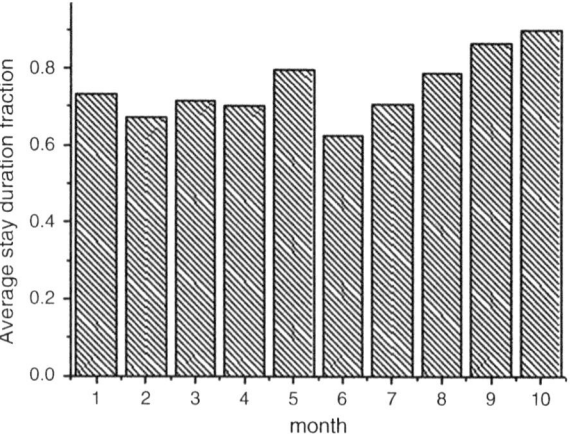

2.1.2 Opportunistic Data Sensing

2.1.2.1 Challenges for Opportunistic Sensing

Opportunistic sensing (OS) refers to a paradigm for signal and information processing in which a network of sensing systems can automatically discover and select sensor platforms based on an operational scenario, determine the appropriate set of features and optimal means for data collection based on these features, obtain missing information by querying resources available, and use appropriate methods to fuse the data, resulting in an adaptive network that automatically finds scenario-dependent, objective-driven opportunities with optimized performance. Theory and algorithms of OS are needed for advancing autonomous sensing that not only ensures effective utilization of sensing assets but also provides robust optimal performance [8].

The wireless medium is often called a fading channel: the pejorative adjective suggests that the intrinsic temporal and frequency variations are an impediment to reliable communication. While not untrue, the channel fluctuations are turned from foe to friend in some scenarios. A concrete situation is when the time scale of communication is much larger than that of the channel fluctuations: the so-called ergodic fading channel. In a point-to-point ergodic fading channel, the transmitter can make good use of the channel state information (CSI): by devoting more power to when the channel is good and less (or even none) when the channel is bad, the rate of reliable communication is improved. The improvement is significant when the operating signal-to-noise ratio (SNR) is small; this is a simple instance of opportunistic communication. Another important instance is in cellular communication: by scheduling user transmissions when their channel conditions are good, the system throughput is improved. This effect is called multiuser diversity [9].

The vehicular sensing network is one special case of opportunistic network used for sensing, which has been envisioned to be used in road safety, emergency responses, and many intelligent transportation system based applications with the rapid growth of city transports [10]. Multi-hop data delivery through vehicular networks is complicated because of the vehicle mobility and varied network environment like topology and traffic, etc. According to this, vehicular data is collected by opportunistic communications, i.e., a vehicle carries the packet until a relay vehicle is in its communication range and then help to forward it in delay-tolerant routing mechanism [11]. Delay-tolerant networks are suitable for applications of vehicular networks [12]. Vehicular DTN can provide a large-scale communications by leveraging the large data storage and energy efficiency of vehicles. The application examples include the use of vehicular DTN to provide low-cost digital communication to remote villages [13] and vehicular sensing platforms for urban monitoring, such as CarTel [14]. But the vehicular DTN network topology changes dynamically, because the network lacks continuous connectivity and may become partitioned at any instant. Then, the uncertainty of network environment in vehicle DTNs is a result of the mobility, limited wireless radio range, sparsity of mobile nodes, energy resources, etc. [15].

2.1.2.2 System Models

We assume that: each node gets knowledge of its locations and is aware of its trajectory by equipping with GPS and pre-loaded digital maps. The node information of its vicinity can be known by beacon messages. There are some static access points deployed in DTN network that help to make connections and disseminate some network routing information. Through the above, each node knows its own location, the location of the destination, and the locations of its potential next hops. In our assumptions, no further traffic statistic of the network is needed.

In order to discuss the routing and scheduling problem, we need some definitions and models to describe the network. A vehicular DTN is composed of vehicles participating in the network with computing and communication capabilities. The links of the network may go up and down over time due to mobility and events, etc. In such DTNs, the routing is based on opportunistic contact of vehicles. More than one contact may be available between a pair of vehicles. In this case, routing metrics are used to make decisions. The contact schedule is the set of time slots that are available for communications, which is related with communication performance. The routing strategies are used to deliver the packets over each hop by knowledge of the networks. The packets will be buffered at each intermediate node because of intermittent network, which enables packets to wait until the next hop is available.

Usually, each node can be in one of the three working states: listening, send and receive data. We assume the system communication is in continuous time period called frames, which can be divided into time slots. DRSS helps to choose the relay and schedule for current transmissions. If no events occur on current node, it can tune into sleep state to save energy.

The energy consumption of DTN nodes includes energy consumed for transmitting E_t, receiving E_r, and listening E_l. Here, we omit the energy consumption of sleep state. The energy consumption can be calculated as Eq. (2.15), where e_e is the energy consumed by transceivers per second, and e_a is the energy consumed in the transmitter RF amplifier per second; e_p is the energy consumed for processing in the receivers, and e_l is the energy consumed for listing to the radio environment. e_l equals e_e. Those parameters are determined by the design characteristics of transceivers. R is the transmission range; n is the power index of the channel path loss. T_t is the time for sending data. T_r is the time for receiving. T_l is the listening time. T is the time length of a cycle.

$$E_t = (e_e + e_a R^n) T_t$$
$$E_r = (e_e + e_p) T_r \qquad (2.15)$$
$$E_l = e_l T_l = e_l (T - T_l - T_r)$$

We assume that communication time is divided into continuous equal frames, which can be divided into continuous equal slots. Each node has the same frame structure and chooses transmit rate according to channel congestion and energy efficiency. The maximum rate communication of node u can be determined from Eq. (2.16). In the equation, C is the channel capacity, W is the bandwidth, P is the transmission power, h is the channel gain, N_0 is the noise power spectrum density, and F is the gap to ergodic channel capacity.

$$\text{rate}_t(u) = \frac{T_c}{T}C = \frac{T_c}{T}W \log_2 \left(1 + \frac{Ph}{N_0 W F} \right) \tag{2.16}$$

In which, T is the time of a frame. T_c is the opportunistic contacting time of two nodes in a frame, where Tc is related with contacting topology changes and rate constraints. It is obvious that $\frac{T_c}{T}$ is proportional to the schedule slot numbers.

The sending and receiving rate on node u can be formulated as:

$$\text{rate}(u) = \text{rate}_t(u) + \text{rate}_r(u) = \frac{T_t + T_r}{T}C \triangleq (a_t + a_r) * C \tag{2.17}$$

In which, $\frac{(T_t+T_r)}{T}$ is the active fraction of slot allocation on current node for sending and receiving data. There is: $T_t + T_r = T_c$. We use λ_t and λ_r to represent the slot allocated fraction for sending and receiving data, respectively. Considering a half-duplex network interface card for node-to-node communications, there is:

$$\lambda_t(u) + \lambda_r(u) \le 1 \tag{2.18}$$

Considering the congestions, for slot τ, it should satisfy Eq. (2.19), where u' is in the interference node set $I(u)$ of node u:

$$\lambda_t(u) + \sum_{u' \in I(u)} \le 1 \tag{2.19}$$

In protocol interference model, transmission methods are utilized to evaluate the interference impact. A node receiving from a neighbor should be spatially separated from any other transmitter by at least a distance D, i.e., interference range. If distance between u and u is less than D, then the two nodes interference with the transmission.

Each node has a buffer to store and carry the packets that have not been transmitted. The node keeps the storage and bandwidth maximally utilized, dropping only when necessary. The buffer has limited maximal size and can contain packets with its available space, e.g., node u has available buffer size as buf(u). Let occ(u, τ) is the number of bytes stored at node u at slot τ under current routing strategies. For all u, τ, there is:

$$\text{occ}(u, \tau) \le \text{buf}(u) \tag{2.20}$$

A delay bound $T_{\max}$ is given to limit the maximal delivery time for data packets. For each node u of the route path with the hth hop in the network, the distance of itself from the destination dst can be measured by Euclid distance as: $d = \|u - \text{dst}\|$. Note that dst can be an average node or an access point. So an approximate average delivery velocity value vector $\bar{v}$ can be calculated as, where the direction of $\bar{v}$ is the direction of node-destination pair:

$$\bar{v} = \frac{\vec{d}}{\text{slack}} = \frac{\|u - \text{dst}\|}{T_{\max} - \sum_{i=1}^{h} T_i} \tag{2.21}$$

In which, slack is the time left for routing, which is equal to the remaining part of delay bound minus cumulative time on each hop for the packet. The velocity vector direction is the direction of current node toward destination.

For the $h+1$ hop, the velocity v of relay node should satisfy, in which θ is the angle of current velocity with the node-destination pair direction:

$$\Delta v = \vec{v} \cos\theta - \vec{\bar{v}} \geq 0 \tag{2.22}$$

Figure 2.9 shows an illustration of, and v for the possible relay of node u, when geographic routing is from s to d. Each node u is assigned with a routing indicator $I(u, u)$ for the next hop u with the data packet, which is defined as cosine value of the angle of current velocity with the node-destination pair direction, i.e., $\cos\theta$. A bigger $I(u, u)$ value indicates a higher delivery priority of the relay.

According to mobility of vehicles, we can find three basic relay patterns in vehicular DTNs as: straight way, intersection toward destination, and intersection against destination. We use: straightway, intersec-2-des, and intersec-against-des to represent the three patterns. If a node carries the packet and its next hop is in the same direction with it toward the destination, then we find out the packet is delivered through vehicles on a straight way that does not change the driving direction either. In this case, we have: $\cos\theta = 1$. If we have $\cos\theta \leq 0$, it implies that the next hop is against from the direction of current node and its destination pair, i.e., *intersec-against-des* pattern. If we have $\cos\theta > 0$ and $\cos\theta < 1$, it implies the next hop is toward the direction of current node and the destination pair, i.e., intersec-2-des

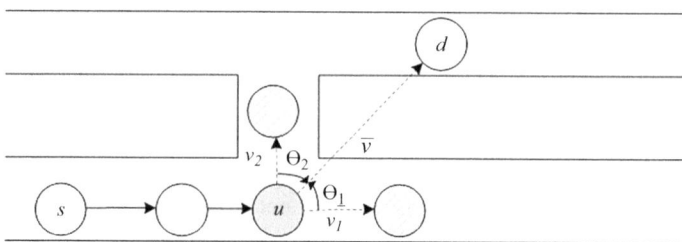

Fig. 2.9 An example for geographic routing

mode. Equation (2.23) gives the definition of the three relay patterns.

$$\text{relay-pattern} = \begin{cases} \text{straightway,} & \text{if } I(u, u') = \cos\theta = 1 \\ \text{intersec-2-des,} & \text{if } I(u, u') = \cos\theta \in (0, 1) \\ \text{intersec-against-des,} & \text{if } I(u, u') = \cos\theta \in [-1, 0] \end{cases} \quad (2.23)$$

In geographic routing, only node moving toward the preferred direction within the delivery latency is probably selected as a relay. But for geographic routing problems in vehicular DTNs, the optimal node may be not connected to destination, so the routing decision only based on velocity policy may make packets be routed to the wrong direction that often leads to higher energy consumption. As shown in Fig. 2.9, if packet on u is forwarded in horizontal direction, the packet will not be successfully delivered because horizontal relay node is not connected toward the destination. Though the connectivity related traffic pattern can be learned by Q-learning, it needs more time and waster energy to do this.

To solve the problem, we combine two routing modes: probabilistic forwarding and replication together in vehicular DTNs to try to speed up the learning process. Which mode to choose is dependent on the packet pattern learned from the network environment. If the current packet is in intersec-against-des pattern, then the packet will not be delivered to the next hop. The node just carries the packet and waits for the other contact. Otherwise, if the packet pattern is straightway, the node will forward the packet to its possible next hop. If the packet pattern is intersec-2-des, the replication method is used to replicate the packet and deliver to the next hop. The node will replicate the packet according to routing indicator, i.e., replicate the packet and deliver it to the next hop with $I \in (0, 1)$.

A feasible routing and rate allocation strategy set $a(\text{relay}(u), \text{rate}(u))$ between routing pair src and dst with maximal delay bound $T_{\max}$ should satisfy:

- **Objective**: The objective is to minimize the total routing and rate allocation cost on the route path, i.e., to minimize the cost of cumulative hops h on the route path as shown in Eq. (2.24).

$$\check{R}(S) = \sum_{1}^{h} \check{R}(u) \quad (2.24)$$

- **Schedule constraints**: The corresponding rate on the path route should satisfy the interference constraint in Eqs. (2.18) and (2.19).
- **Buffer constraints**: The relay node on the path should satisfy the buffer requirement in Eq. (2.20).
- **Delay constraints**: The relay node on the path should satisfy the delay requirement in Eq. (2.22).

Fig. 2.10 The intelligent
network control framework

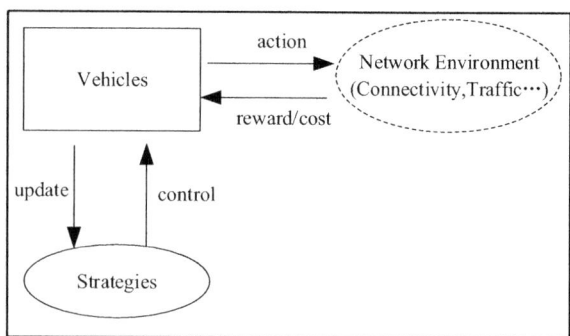

2.1.2.3   DRSS Design

Network Control for Green Opportunistic Communications

The urban vehicular network traffic changes are related with time, zone vehicle
types, and some other factors. Vehicles follow a kind of certain mobility pattern
that can be a combination of the underlying roads, the traffic lights, the speed limit,
traffic condition, etc. We try to learn from these traffic patterns and make it to assist
data delivery in vehicular DTNs. We design a directional routing that helps to relay,
carry, and forward the packets by selecting the routes within the preferred areas that
may be with high vehicle density and more preferred vehicle toward the destination.
An intelligent network control based on reinforcement learning is used to predict
the knowledge of the network environment. The probabilistic routing strategies are
updated according to reword/cost from the current network environment. We know
that directional routing problem in vehicular DTNs complicates for node mobility.
Reinforcement learning based network control framework can help to predict the
motion direction and network traffic pattern according to history experiences. The
probabilistic routing strategies converge toward a best route of a specific direction.

   Figure 2.10 shows the network control framework by using reinforcement
learning. To deal with the opportunistic connectivity environment in DTNs, the
data communication should be based on the prediction of future contacts by taking
advantages of nodes' mobility history. During the intelligent learning procedure,
the vehicle adapts its strategy according to its reward or cost with different actions
without known modeling of network environment. With reinforcement learning,
the vehicle can update its strategy by predicting the network environment and
converging to the optimal strategy.

   We adopt Q-learning to circumvent the routing and scheduling problems in
DTNs. Q-learning is used for learning and decision under dynamic and unknown
environment model. The DTN network system has random connection events
because of mobility. The Q-learning based routing scheme observes the transmission
activity of its nearby nodes and gathers the information of network traffic and
topology as well as route resource utilization, which enables the node to build the

knowledge of nearby routing environment. Then, the knowledge is used by the vehicle node to decide which relay direction and schedule to choose so that the desired performance can be achieved. Provided the learned information is accurate, it allows routing and schedule strategies to make efficient use of network resources by routing packets toward optimal direction. The objective of DRSS is to learn from the environment states (dynamic connection events) to decide on actions so as to maximize the reward (i.e., minimize the cost function) [16]. The DTN control system is formulated by a tuple $\langle S, A, \check{R}, \check{T} \rangle$, where S is the discrete hazard state space. A is the discrete action space that is dependent on strategies taken. $\check{R} : S \times A \rightarrow \mathbf{R}$ is the cost function, which implies the quality of a state-action combination of the network system. $\check{T} : S \times A \rightarrow \Delta S$ is the state transition function, where ΔS is the probability distribution over state space S. In DRSS, the actions include sets of selected relay and rate strategies, i.e., a set of tuple pairs as: $(\text{relay}(u), \text{rate}(u))$, in which $\text{relay}(u)$ is the directional localization for the relay, and $\text{rate}(u)$ is the assigned slots for the communication. Once an action (i.e., strategy of the DRSS scheme) is taken, the network system produces new performance signal (i.e., cost) according to it. Then DRSS receives the update cost $\check{R}$, which is used to evaluate the effectiveness of the action. The learning procedure is achieved by updating the Q-value. The Q-learning approach converges to an optimal strategy as long as the state-action pairs are continually updated. When traffic relay transmission beginning at time slot τ is finished at the next time $\tau + \lambda$ (among it, λ is the time slots for transmission schedule), then the Q-value for state-strategy pair is updated by Eq. (2.25).

$$Q'(s, a) = (1 - \alpha)Q(s, a) + \alpha(\check{R}_\tau + \beta \min_a Q(s', a')) \qquad (2.25)$$

In Formula (2.24), α is the learning rate and in the range of $(0,1)$. β is the discount factor and is in the range of $(0,1)$ too. We use a constant learning factor, and the learning procedure can track the dynamic network situations.

The DRSS scheme aims to provide green energy-efficient routing and rate allocation. The cost function is achieved by the average amount of energy consumption per bit on the current route path from 1 to h hops, which embodies the trade-off among energy efficiency, connection duration, and communication efficiency:

$$
\begin{aligned}
\check{R}(\text{path}) &= \sum_1^h \check{R}(u) = \sum_1^h \frac{E}{\text{rate}(u) * T_c} = \sum_1^h \frac{E_t + E_r + E_l}{\text{rate}(u) * T_c} \\
&= \sum_1^h \frac{(e_e + e_a R^n)\lambda_t + (e_e + e_p)\lambda_r + e_l(1 - \lambda_t - \lambda_r)}{(\lambda_t + \lambda_r)C * (\lambda_t + \lambda_r)}
\end{aligned}
\qquad (2.26)
$$

Considering the flow constraints and buffer constraints, we formally express the cost function of DRSS strategy S as Eq. (2.27).

$$\check{R}(u) = \begin{cases} \frac{(e_e + e_a R^n)\lambda_t + (e_e + e_p)\lambda_r + e_l(1 - \lambda_t - \lambda_r)}{(\lambda_t + \lambda_r)C*(\lambda_t + \lambda_r)}, & \text{if satisfies (4)-(8)} \\ \infty, & \text{otherwise} \end{cases} \tag{2.27}$$

For one-commodity routing situation, if node $u \notin src, dst$ on the route, then the traffic data bits sent out are equal to the received data bits, i.e., $\lambda_t = \lambda_r$. The received traffic rate depends on the traffic rate sent out of the last hop on the route. So, the DRSS scheme is only responsible for relay selection and allocating transmission slots.

For online multi-commodity cases, Q-learning approach is extended for multi-agent decision-making. In online multi-commodity situations, each agent is selfish for routing desire. The network routing and allocation process is modeled as a stochastic non-cooperative game framework. Let $\langle N, S, \check{R} \rangle$ denote the stochastic non-cooperative routing and scheduling game, where N is the number of routing commodities in the network, S is the strategy set, and $\check{R}$ is the cost set. The objective of multi-commodity routing is to minimize the overall cost. Different agents have their own objectives which could be conflicting with each other, so the overall cost is dependent on the strategy selection of each agent. The strategy of each agent is responsible for selecting routes and rates (i.e., transmitting slots) that the overall multi-commodity routing strategy cost achieves minimized cost during the learning process. Considering the multi-commodity situation, multi-agent Q-learning approach is utilized to find out the optimized DRSS strategy to achieve the minimized cost for all route paths. We consider $\pi_i(S)$ as the probability for ith routing commodity to select a specific strategy S at time τ, which achieves the game equilibrium. Π_i is the set of possible strategies. The overall objective in the stochastic non-cooperative routing and scheduling game can be expressed as:

$$\min_{\pi_i \in \Pi_i} E[\check{R}_i(S)], \quad \forall i \tag{2.28}$$

where

$$E[\check{R}_i(S)] = \sum_i \pi_i(S)\check{R}_i(S) \tag{2.29}$$

In the stochastic non-cooperative game based network system, each routing commodity finds a strategy with Nash equilibrium to achieve objective in Formula (2.28). Let $\text{Nash}Q_i$ be agent's cost in current state with the selected equilibrium. According to [17], the Nash equilibrium $\text{Nash}Q = (\pi_1^*, \pi_n^*)$ is computed from Formula (2.30). In the formula, m is the number for players in the game. In order to calculate the Nash equilibrium, each agent i need to know the other agents' Q-values. Then, each agent observes the other agents' immediate costs and actions. So, agent i can update its Q-value according to other agents' Q-values as shown in Eq. (2.31). In each time step at time τ, a player observes the current state s, and takes action a. An immediate cost $\check{R}$ and the next state s are observed.

$$\text{Nash}Q_i(s,a) = \sum_1^m \pi_i^*(s,a)Q_i(s,a) \tag{2.30}$$

$$
\begin{aligned}
Q_i'(s,a_1,\ldots\ldots,a_m) &= (1-\alpha)Q_i(s,a_1,\ldots\ldots,a_m) \\
&\quad + \alpha(\check{R}_i + \beta \text{Nash}Q_i(s',a_1,\ldots\ldots,a_m))
\end{aligned}
\tag{2.31}
$$

To minimize the network cost, the multi-agent Q-learning approach has to explore all possible strategies randomly and greedily, and then chooses the good strategy. The strategy exploration probability is updated as shown in Eq. (2.32). γ is a constant factor between 0 and 1. The learning policy satisfies the GLIE (greedy in the limit with infinite exploration) property and also based on routing indicator priority. The relay is based on the priority of routing indicator I, and the bigger indicator value means higher priority.

$$
\pi_i'(a^*) = \begin{cases}
\pi_i(a) + I(1 - \pi_i(a)), & \text{if } a^* = \arg\text{Nash}Q_i(s,a) \\
\max(I\pi_i(a), 0), & \text{if } a^* \neq \arg\text{Nash}Q_i(s,a)
\end{cases}
\tag{2.32}
$$

The Nash Q-learning based DRSS algorithm is described in Algorithm 1. Line 1–6 is the algorithm initialization. In line 4, $|A|$ is the number of possible strategies, which is bounded by the multiplication of maximal degree of the network graph Δ and frame slot number L, i.e., $|A| = \Delta * L$. Line 7–27 is the Nash Q-learning procedure. Line 13–19 is the routing mode selection, i.e., probabilistic forwarding, directional replication, and carry. In line 20, $\check{R}_1 \check{R}_m$ donates the cost for all players, and a1 am donates the strategy taken by the other players except ai. In line 21, slot time τ is updated as $(\tau + \lambda) \bmod L$, where L is the frame length. Line 23 shows the Q-value update of each user for its next state according to Eq. (2.31).

As explained in [17], the time complexity and space requirement of this learning algorithm are high when agent number is big. For 2-player Nash Q-learning, it has exponential worst-case time complexity. The space complexity is also exponential in the number of users. In the network, the game of routing resources occurs among the routing commodities within the interference range during the contacting time. The transmissions diverged from interference range will not affect one another. So, the routing game process of the network system can be achieved by the local game with local routing commodities, when there are joint routes within the interference range from one another during the contacting time.

An improved DRSS algorithm based on local games is illustrated in Algorithm 2. In Algorithm 2, we only consider the local game that happens among the interference range to reduce the complexity. In line 21, j is the routing commodity with routes that are within the interference range of i for current contacting state by satisfying: $\|\text{route}(i) - \text{route}(j)\| \leq D$. D is the interference range. The distance of the routes is defined as the minimal distances of the relay nodes from the two routes. So, we only observe the cost and action of contending

---

Algorithm 1 Basic DRSS algorithm

1: **for** $i=1\ldots m$ //m agents with src_i and dst_i **do**
2:     Let $\tau=0$, get the initial state $s = s(\tau)$
3:     **for** all $s \in S, a \in A$
4:   **do** $Q_i(s, a^1, \ldots, a^m) = 0, \pi_0^i(a) = 1/|A|$
5:     **end for**
6: **end for**
7: **while** (network execution condition is TRUE) **do**
8:     $u_i = \text{src}_i$
9:     **for** current node u_i //u_i is the current hop for i^{th} agent **do**
10:         **if** $(u_i == \text{dst}_i)$ **then** break;
11:         **end if**
12:         Choose action a_i according to (18)
13:         **if** $\text{relay}_{\text{pattern}}==$ straightway **then**
14:             forward the packet; //forward one copy
15:         **else if** $\text{relay}_{\text{pattern}}==$ intersec-2-des **then**
16:             replicate the packet; //replicate the packet
17:         **else if** $\text{relay}_{\text{pattern}}==$ intersec-against-des **then**
18:             break; //carry the packet until next contact
19:         **end if**
20:         Take action a_i for $s(\tau)$ and observe $\check{R}_1 \check{R}_m, a_1 a_m$
21:         update $\tau = (\tau + \lambda) \bmod L$, the next state $s = s(\tau)$
22:         **for** all $j \neq i$, j$=1\ldots m$ **do**
23:             Update Q_j according to (17)
24:         **end for**
25:         $s = s, u_i = a(\text{relay}(u_i))$//update state and the next hop
26:     **end for**
27: **end while**

---

commodity with current commodity i, but not all the other commodity in the network. For classical DTN networks and their applications, the network is usually sparsely or loosely connected. The online routing commodity number is often small within the interference range, so the performance of local game based DRSS is acceptable. From the above algorithms, DRSS selects the contacts moving toward the energy-efficient and interference-aware direction toward the destination within the timeliness by feasible schedule. According to Q-learning cost, DRSS converges to select relay nodes that are moving with velocity equal to or higher than the expected average velocity toward the destination direction with good connectivity and traffic situations. Otherwise, the node carries the packets until it finds a better node that can help to make data delivery.

Next make simulations by using ns2 simulator and evaluate the performance of DRSS. We simulate a vehicular ad hoc delay-tolerant network with 50 vehicles around 2500 m × 2500 m area. The vehicles travel with speed from 5 to maximal 20 m/s around the test area. A predefined roadmap and moving model based on random waypoint is given for each node. A fixed static node is placed at the center of the area, which is used to simulate an access point in real vehicular networks. By default, all the vehicle nodes move in the network according to determined mobility model. In the simulations, the vehicles connect with each other by a half-

---

Algorithm 2 Improved DRSS based on local game

---

for $i = 1 \ldots m$ //m agents with src$_i$ and dst$_i$ **do**
    let $\tau=0$, get the initial state $s = s(\tau)$
    for all $s \in S, a \in A$ **do** $Q_i(s, a^1, \ldots, a^m) = 0, \pi_0^i(a) = 1/|A|$
    end for
end for
while (network execution condition is TRUE) **do**
    $u_i = \text{src}_i$
    for current node u_i //u_i is the current hop for i^{th} agent **do**
        if (**then**$u_i == \text{dst}_i$) break;
        end if
        choose action ai according to (18)
        if relay$_\text{pattern}$==straightway **then**
            forward the packet; //forward just one copy
        else if relay$_\text{pattern}$==intersec-2-des **then**
            replicate the packet; //replicate the packet
        else if relay$_\text{pattern}$== intersec-against-des **then**
            break; //carry the packet until next contact
        end if
        take action ai for $s(\tau)$ and get $\check{R}_i$
        observe $\check{R}_j, a_j (j \neq i, j = 1m), if \|\text{route}(i) - \text{route}(j)\| \leq D$
        update $\tau = \tau + \lambda$, the next state $s = s(\tau)$
        for a **doll** j //j donates the local contender
            Update Q_j according to (17)
        end for
        $s = s, u_i = a(\text{relay}(u_i))$//update state and next hop
    end for
end while

---

duplex wireless network interface card. The communication between access point and vehicles uses a separate network interface card to prevent interference. The communication range is 150 m, while the interference range is set to 300 m. At the beginning of the simulation, the routing request with CBR traffic is periodically generated from a source node that is randomly selected. The traffic load is 40 Bps generated on the source node. By default, each node allocates a limited buffer with maximal 30 packets in buffer. One slot is enough time to transmit or receive a packet. The simulated access point is fixed chosen as routing destination. We initialize the parameter values: $\alpha = 0.1$, $\beta = 0.9$, and $\gamma = 0.5$. The value of energy consumption related parameter e_e, e_a, e_p, and el is chosen based on [18].

We simulate and evaluate the convergence of Q-learning based DRSS, learning ability when traffic load changes, energy efficiency, and then the performance of data delivery delay. We make simulations when the routing commodity number is 1, 2, and 4. In 4-commodity situations, we use local game method based algorithm.

DRSS utilizes Q-learning approach, so convergence is one of the main concerns for the algorithm performance. We study the convergence speed of the cost function as simulation time increases. The result is shown in Fig. 2.11. It shows the average path cost from 20 to 1000 s during the simulation. We observe that the average path cost begins to converge around 800 s for all the situations. When comparing multi-

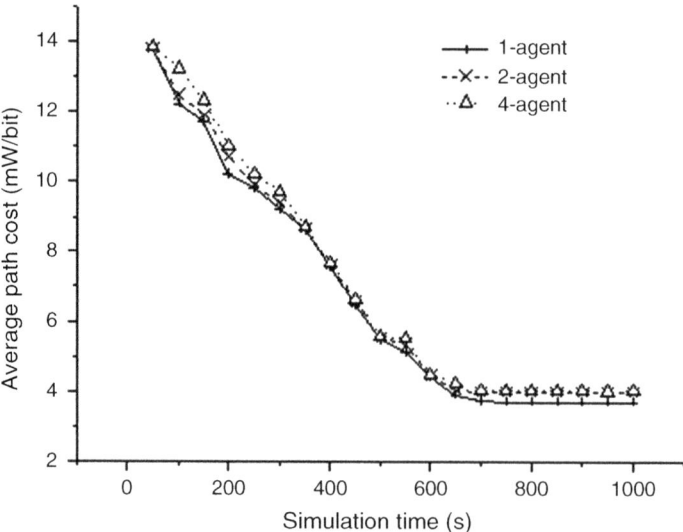

Fig. 2.11 Convergence of DRSS

agent results with 1-agent situation, they seem to have a little more path cost. This is because Nash equilibrium strategy is chosen to provide optimal routing cost for all the agents in the network, which may not be the optimal strategy for each single agent.

The advantage of Q-learning approach based DRSS is the ability to predict the unknown network environment, learn with it and then adapt the strategies toward it. In practical DTNs, network topology and traffic change with time. We then make simulations with the varied network scenarios.

The first scenario is: we add 20 more default vehicles into the network to help the connections at 200 s after the simulation begins, and then we remove them at 800 s. The corresponding results are shown in Fig. 2.12. We observe that DRSS can track the change and adapt the strategies toward the network environment. The average path cost of routing situations with 20 more vehicles added into the network is lower than the original situations. It implies that the added vehicles help to improve the contacting opportunity in the network without extra congestions because of effective rate allocation control. At 800 s of the simulations, we remove those added vehicles. It incurs small fluctuations of average path cost, because each agent has to predict and learn new strategy toward the changes.

The second scenario is: we double the source traffic load from 200 s of the simulations, and then recover to original at 800 s. The corresponding results are shown in Fig. 2.13. We observe that the average path cost increases a little from 200 s to 800 s when compared with the previous results. This is because the increased traffic load incurs possible congestions. According to DRSS, the routing and rate strategies will adapt to improve the cost. Then more idle energy consumption is

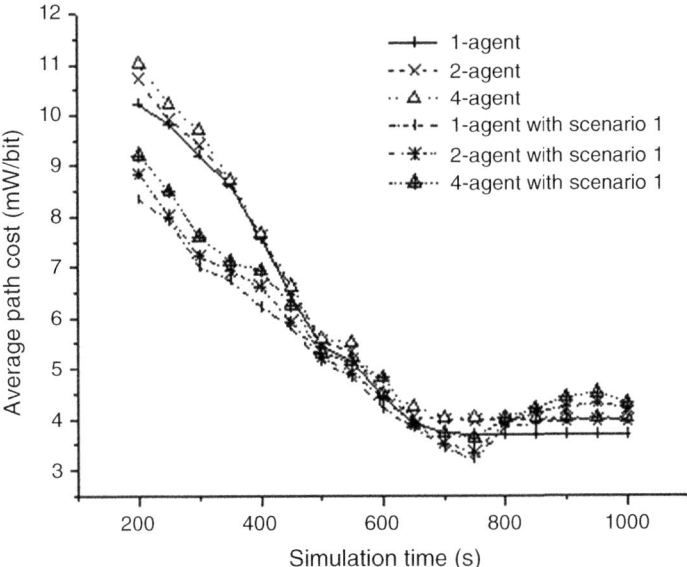

Fig. 2.12 Learning ability of DRSS with scenario 1

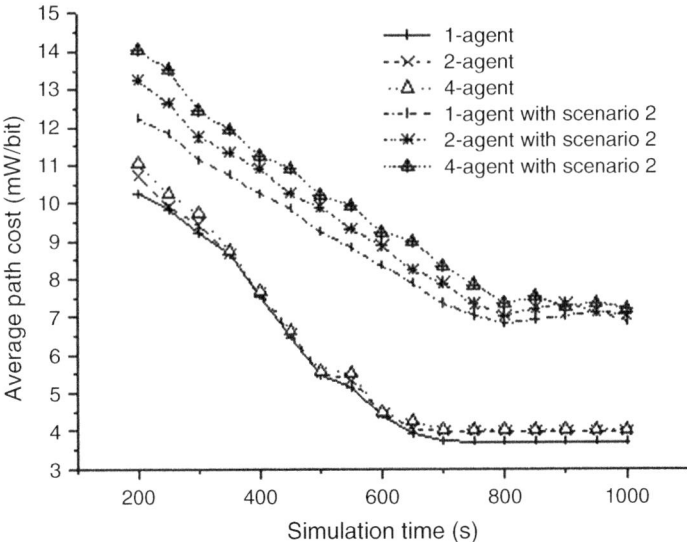

Fig. 2.13 Learning ability of DRSS with scenario 2

needed in long schedule for the increased traffic. The results show there are many fluctuations after 800 s, because DRSS learns and select new strategies toward the current the new network environment.

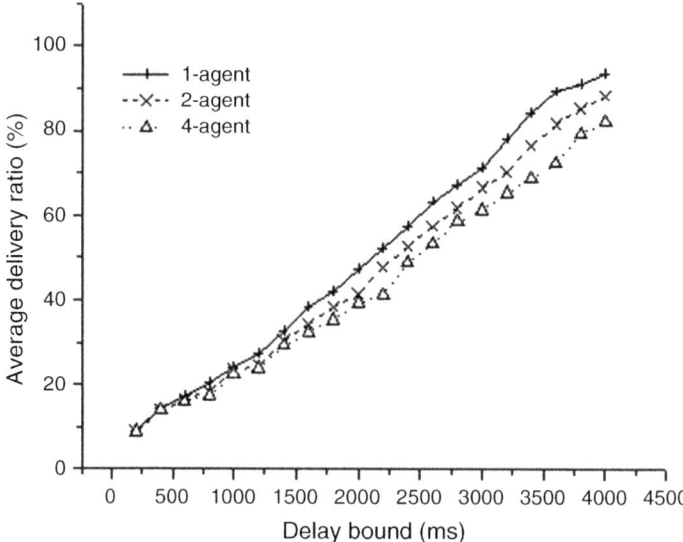

Fig. 2.14 Data delivery ratio as delay bound increases

Another concern of DRSS performance is the data delivery ratio within the given delay bound. Figure 2.14 shows the data average real-time delivery ratio as delay bound increases from 200 to 4000 ms, which is based on results of 800 s simulations. We observe that the data delivery ratio increases as we lose the delay bound. And 1-agent achieves better delivery when compared with 2-agent and 4-agent situations. This is because more routing agents bring more possible congestions in the networks and need more schedule time for transmission according to DRSS rate allocation.

We compare our DRSS with epidemic [19] and modified Dijkstra algorithm [20]. Figure 2.15 shows the average energy consumption per bit during the simulations. From the results, we can see that DRSS has much better energy efficiency when compared with the other two algorithms. Epidemic algorithm uses replication to deliver data packets in the network, so more energy is needed. And this is even worse when simulation goes on, because much replicated routing occurs during the packet delivery. The modified Dijkstra algorithm uses forward method to minimize the average delivery delay, but it does not consider the energy consumption of packets and cannot adapt toward the varied network environment. Actually, sometimes it leads to the wrong direction that is close to the destination without connection and may consume more energy.

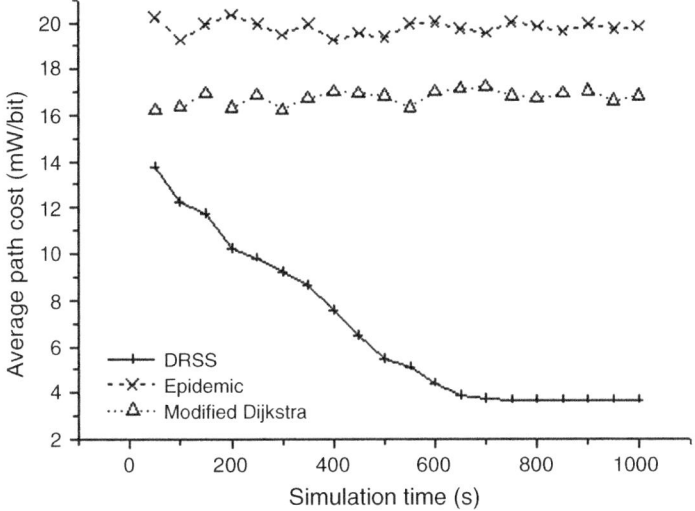

Fig. 2.15 Average energy consumption per bit

2.2   Feature Discovery of Mobile Data Users and Networks

2.2.1   Central Behavior Based Feature Discovery

2.2.1.1   Characteristics of Mobile Internet Usage

In this section, we study the characteristics of mobile network usage on a large-scale usage detail records (UDRs), which is described in detail in Sect. 2.2.1.3. First, we investigate the diversity of mobile internet users on spatiality and interest. Then we measure how predictable are mobile users' interests and the effect of spatial–temporal information on predicting them. Finally, this paper introduces centrality phenomenon both on spatiality and temporality which inspires our feature transformation method.

First of all, we take an overview of the diversity of mobile users. The diversity of a user means the number of unique interests/locations visited by him. The statistical results are shown in Fig. 2.16.

Both the diversity of location and interest show the form of lognormal distribution, which is widely found in the analysis of human behavior [21, 22]. Users with extreme location/interest diversity are rare, and most of the users have a constrained scope both on spatiality and interest. Specifically, users are more limited on spatiality than interest. The distribution of interest diversity is wider and shorter than that of location diversity. Since it is much easier for users to explore a new website than implementing physical movement, users are more free and willing to explore new interests on internet, which makes interest prediction much trickier and more valuable. We then measure the predictability of mobile internet users.

Fig. 2.16 The diversity of mobile users' interests/locations. The horizontal axis represents the level of corresponding diversity, while the vertical axis denotes proportion of corresponding users

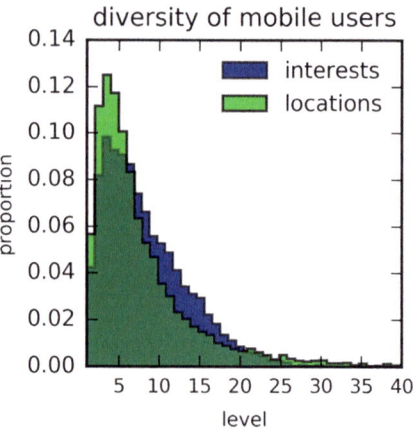

To evaluate the predictability of mobile users' interests, this paper utilizes information entropy inspired by Chaoming et al. [23]. A larger entropy value means the larger uncertainty. First, the max (or random) entropy of user i is defined as

$$H_i^{\max} \equiv \log_2 k_i \qquad (2.33)$$

where k_i is the number of unique interests in the whole observation period. $H_i^{\max}$ indicates the maximum randomness of user i. Then the uncorrelated entropy is

$$H_i^{\mathrm{un}} \equiv - \sum_{j \in N_i} P_i(j) \log_2 P_i(j) \qquad (2.34)$$

where N_i is the set of interests containing k_i elements, and $P_i(j)$ is the probability of activity j. H_i^{un} indicates the predictability of user i. And finally, the conditional entropy of user i is defined as

$$H_i^{\mathrm{con}} \equiv - \sum_{j \in M_i} P_i(j) \sum_{l \in N_i} P_i(l \mid j) \log_2 P_i(l \mid j) \qquad (2.35)$$

where M_i is the extra information set with m_i elements. H_i^{con} indicates the predictability of user interest when an auxiliary feature is specified. We collect the three kinds of entropy mentioned above for each user, and give the cumulative distribution function in Fig. 2.17. The results show that mobile users' interests are far from random, on the contrary, they can be predicted to a certain degree. Moreover, both temporal (hour) and spatial (location) information contribute to improving predictability, which inspires us that both spatial and temporal information are meaningful and helpful. A natural question is why spatial and temporal information make mobile users' interests more predictable, and we will discuss it on spatiality and temporality in the following sections.

Fig. 2.17 The cumulative distribution of users with max, uncorrelated and conditional entropy. The form A| B means the conditional entropy of A under the condition B. To obtain the conditional entropy of activity under the condition time, we extract the hour in timestamp from each record

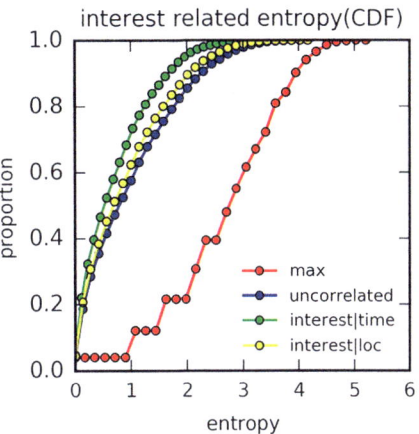

To investigate the relation of mobile network usage on spatiality, we refer to tie strength theory proposed in [24]. Let $S_i = s_i^1, s_i^2, \ldots, s_i^n$ denote the locations that user i visited. Then the strength of tie between user i and base station j is $t_{ij} = (nT_{ij})/(\sum_{k=1}^n T_{ik})$, in which T_{ik} is the total time that user i contributes to base station k. We then define the strong tie when $t_{ij} \geq 1$, and weak tie when $0 < t_{ij} < 1$. For every user, this paper collects his network usage duration at each location and resort them in descending order. We then compute the distances of current record to the top five locations and collect strong and weak ties. The complementary cumulative distribution function (CCDF) is shown in Fig. 2.18a, b. Both strong and weak ties are statistically concentrated in the scope of several most favorable locations. Given a certain frequent visited location, the further a place is apart, the less likely it is to attract mobile network usages. Under certain distance, the probability that network usages occur at a location tends to be in proportion to its popularity. Besides, over 70% of network usages (strong/weak ties) are contributed to the most frequent popular location (location 1), and the second popular location (location 2) attracts about 18% strong and weak ties.

Figure 2.18a, b indicate that users tend to access mobile internet at several locations and they may be more informative compared to traditional spatial features, such as gyration and distance between consecutive records on spatial. Similar to the philosophy of principle components analysis (PCA), we select hotspots defined in Sect. 2.2.1.2 as reference anchors to specify their effects on current online activity on spatiality. The details about this process are presented in Sect. 2.2.1.2.

2.2.1.2   Feature Transformation Based Central Behavior

According to the analysis provided in Sect. 2.2.1.1, mobile users reflect centralities both on spatiality and temporality. Specially, users tend to access mobile internet at several most frequently visited locations (hotspots) on spatiality, and contribute

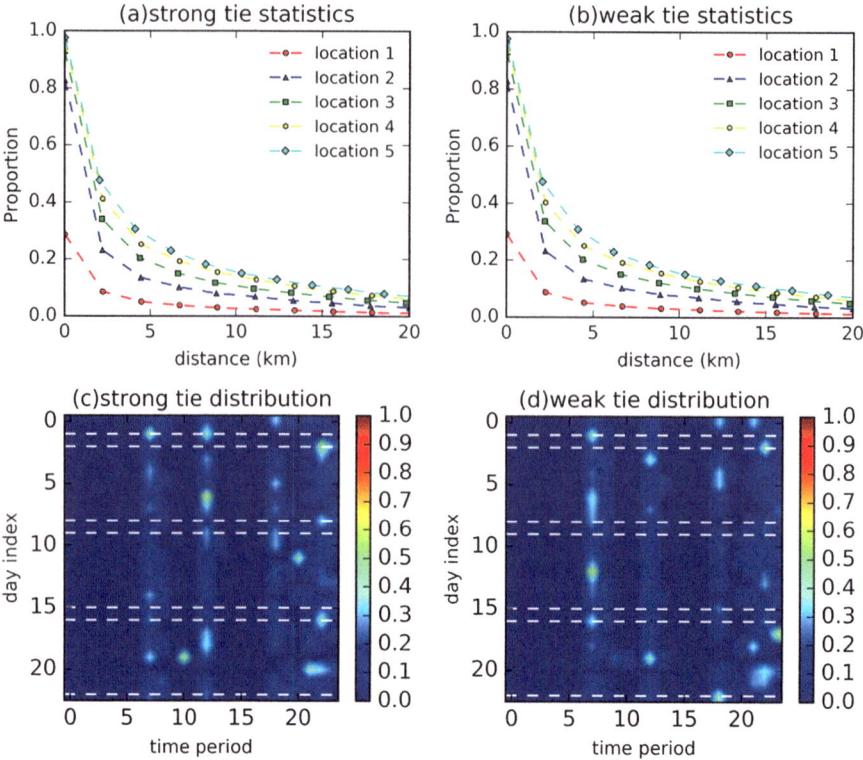

Fig. 2.18 Network usage characteristics on spatiality and temporality. (**a, b**) The complementary cumulative distribution of strong/weak tie from the five most favorable locations. (**c, d**) Network usage of a typical user in observation period (23 days). Color means the proportion of network usage in strong/weak ties. The white dashed lines indicate weekends. Horizontal axis represents timeslot, and the vertical axis is index of days

relatively more network usages at particular timeslots (hot-times). We then integrate these centralities into the designing of TCB.

The intuition behind TCB is that centralities (hotspots and hot-times) can affect user online activity. We assume that these centralities are stationary on time series, and the more closer a centrality is, the influential it can be. For the simplicity of illustration, we make the definition below:

Definition 1 Let hs_m^i be the m-th hotspot of user i, the influence that hs_m^i affects online activity occurred at location j is defined as $\mathrm{HS}(i, j, m) = I(hs_m^i)\exp(-\mathrm{dist}(j, hs_m^i))$. $I(hs_m^i)$ is the total influence of hs_m^i, and $dist(j, hs_m^i)$ is the Euclidean distance between hs_m^i and location j;

Definition 2 Let ht_n^i be the n-th hot-time of user i, the influence that ht_n^i affects online activity occurred at timeslot k is defined as $\mathrm{HT}(i, k, n) =$

Algorithm 3 TCB(records,candidates,win,m,n)

1: # obtain hotspots for every user in win days;
2: hotspots ← CentralityDetection(records,candidates,win,'spatial',m);
3: # obtain hot-times for every user in win days;
4: hottimes ← CentralityDetection(records,candidates,win,'time',n);
5: # obtain statistical information at hotspots and hot-time;
6: SI ← Statis$_{\text{Info}}$(records,hotspots,hottime,win);
7: result=[ ];
8: **for** record in records **do**
9:     u, l, t ← obtain basal information for current record;
10:     # obtain the effect of each hotspot to current location l;
11:     $d_1^u, \ldots, d_m^u$ ←Effect_by_HS(l,hotspots[u]);
12:     # obtain the effect of each hot-time to current time t;
13:     $e_1^u, \ldots, e_n^u$ ← Effect_by_HT(t,hottimes[u]);
14:     temp$_{\text{result}}$ ← concate([$d_1^u, \ldots, d_m^u$], [$e_1^u, \ldots, e_n^u$], SI[u]);
15:     result.append(temp$_{\text{result}}$);
16: **end for**
17: **return** result;

$I(ht_n^i) \exp(-\text{inter}(k, ht_n^i))$. $I(ht_n^i)$ is the total influence of ht_n^i, and $\text{inter}(k, ht_n^i)$ is the time interval between hot-time ht_n^i and k;

Definition 3 For user i, the total influence of the k-th centrality is $I(c_k^i) = (\text{NS}(c_k^i))/((\text{NW}(c_k^i) + 1))$, where $\text{NS}(c_k^i)$ and $\text{NW}(c_k^i)$ are the numbers of strong and weak ties at centrality c_k^i, respectively. The definitions of strong/weak ties are presented in Sect. 2.2.1.1.

Algorithm feature transformation based central behavior (TCB) is presented in the following steps:

To begin with, TCB collects m hotspots and n hot-times for each user, and the results are in descending order according to their influences in win days. CentralityDetection is discussed in detail in the next section. For better behavior description, TCB obtains the statistical summaries at each hotspot and hot-time to describe user behavior. To this end, we refer to [17, 25, 26] and select average displacement to hotspot, average record duration on hotspot, average time interval to hot-time, and average record dwelling time on hot-time to characterize user behavior. In Sect. 2.2.1.3, we will demonstrate how these statistical summaries affect the performance of user interest prediction.

Then, in the loop $L8 - L15$, TCB first fetches original information such as user u, time t, and location l from current record. Then TCB computes the effects that current record received from each hotspot and hot-time according to its influence, namely $d_1(e_1)$ is the effect received from the most influential hotspot (hot-time), and d_m (e_n) is the result of least influential one. The function Effect_by_HS(Effect_by_HT) is designed according to the definitions mentioned above. Note that the number of real hotspots (hot-times) can be less than m (n), we complement the values related to missing hotspots or hot-times by zeros since these hotspots/hot-times have no effect on current record.

Table 2.1 Symbols and corresponding illustration

Original features (O)	dur	Dwelling time of the record
	long	Longitude of the location of the record
	lat	Latitude of the location of the record
	t	Timeslot that record occurred
Temporal features (T)	Δt	Time interval between two consecutive records
	Δt_s	Time interval between two consecutive records in same interest
Spatial features (S)	Δl	Distance interval between two consecutive records
	g	Gyration
Hotspots related features (HS)	d_l^u	12 The effect of hotspot i to current location l for user u
	$dis_{i,aver}^{u,win}$	The average displacement to hotspot i in win days for user u
	$sd_{i,aver}^{u,win}$	The average record duration on hotspot i in win days for user u
Hot-times related features (HT)	e_i^u	The effect of hot-time i to current timeslot t for user u
	$int_{i,aver}^{u,win}$	The average interval to hot-time i in win days for user u
	$td_{i,aver}^{u,win}$	The average record duration on hot-time i in win days for user u

In this way, original spatial–temporal information in raw record is projected into a new vector space, of which coordinates represent the effects received from the centralities (hotspots or hot-times) ranking in certain order based on their influences. Finally, the effects suffered from hotspots/hot-times, together with the statistical information at hotspots and hot-times are concatenated into one record. The results returned by TCB are used for model training and validation.

Based on the description above, all features used in this paper are shown in Table 2.1. They are classified into five groups according to their generation and background. In particular, HS and HT are feature sets produced by TCB. Both HS and HT consist of effects received from centralities and statistical summaries at centralities.

As mentioned in Sect. 2.2.1.1, hotspots and hot-times are behavior centralities on spatiality and temporality, respectively. Therefore, the philosophy behind CentralityDetection is similar to hot points detection in time series. Different from [27], CentralityDetection is designed to find the most influential k centralities specified by matric (spatial or temporal) according to the influence values defined in Sect. 2.2.1.2. When matric is set to spatial, the historical locations of each user are processed, otherwise the timeslots. For the simplicity of illustration, we refer location and timeslot as point. CentralityDetection collects the numbers of strong ties and weak ties at each point. In this case, NS and NW are the vectors with same length. Then the influences of all points are computed according to Definition 3 and stored in dictionary influence by descending order based on the influence values. The number of centralities is the minimum of k and the length of points list. Therefore, for some users, the number of their centralities can be less than k if they use mobile internet at less than k locations or timeslots. The pseudocode of CentralityDetection is presented as follows.

Algorithm 4 CentralityDetection(records,candidates,win,matric,k)

1: r_win $\leftarrow$ obtain records occurred in win
2: result_dic $\leftarrow \phi$ #initialize result dictionary
3: **for** user in candidates **do**
4:     NS $\leftarrow$ collect the number of strong_tie in r_win at each point on matric
5:     NW $\leftarrow$ collect the number of weak_tie in r_win at each point on matric
6:     influence sort (NS/((NW+1))) in descending order
7:     c_list=[ ]
8: **end for**
9: **for** r, i in range(min(len(influence),k)) **do**
10:     c_list.append(influence[i].point)
11: **end for**
12: add user,c_list into result_dic
13: **return** result_dic;

Let N be the number of candidates, and k the number of centralities. Then the complexity of Algorithm 4 is $O(Nk)$. Hotspots or hot-times obtained in this process are further used to collect statistical information, and compute the effects they bring to mobile network usages.

In this section, this chapter investigates the correlation between different feature sets with users' interests. Distance correlation (DC) $R \in [0,1]$ is a new metric to measure the dependence between two random variables. It equals to 0 if and only if the two random variables are independent [28]. DC is effective both in linear and nonlinear situations. Besides, it can be applied between the variables with different dimension, and regardless of whether it is categorical, continuous, or discrete [29]. Thus DC is quite suitable for measuring the correlation between feature sets and the corresponding interests.

To make the distance between different interests computable, this paper utilizes dummy variables to represent each interest. For the details about DC we refer the readers to [28], and we apply the package energy [30] in the process of computing DC values. Since the complexity of DC is $O(n^2)$ (n denotes the number of samples), it is impractical to compute the DC values from an overall perspective. Thus this paper utilizes different steps to split whole records into several blocks and collects the DC values of different feature sets at each block.

As shown in Fig. 2.19, in general, both HS and HT have similar performance, and show great advantages over O/S/T, indicating that feature sets generated by TCB are much more informative in predicting mobile users' interests. The performance of O and S is similar, while T ranks the worst. It also suggests that classical spatial–temporal features (T and S) are limited in predicting mobile users' interests. In Sect. 2.2.1.3, we will compare the performance of different feature sets in detail.

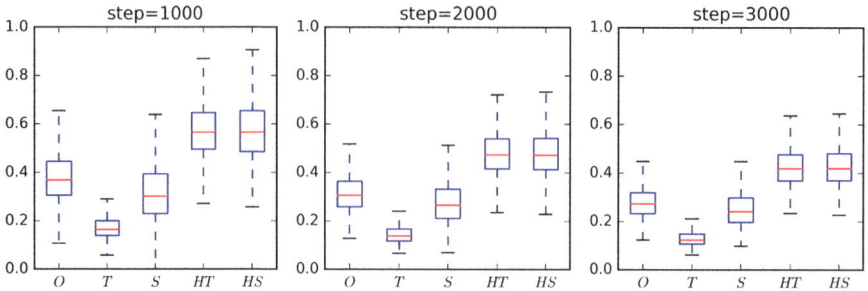

Fig. 2.19 The correlation between different feature sets with users' interest in various steps. The red line in each box indicates median. Only three steps are taken due to the huge time complexity. The numbers of hotspots and hot-times are set to 3, and statistical window is 7

Table 2.2 Key fields and examples in UDRs

PhoneNUM	Start	End	Location	URL
68960814031	2014-11-21 12:02:19	2014-11-21 12:02:26	689B_83A1	m.baidu.com
69061452339	2014-11-21 06:23:03	2014-11-21 06:23:20	67BD_3345	m.sohu.com
69061454535	2014-11-21 13:40:52	2014-11-21 13:41:09	67BD_5BB9	wap.cmread.com

2.2.1.3   Evaluation

In this section, we compare the performances of various feature sets under the state-of-the-art classification algorithms. In particular, we investigate the performance of X/STA to analyze the importance of statistical summaries related to feature set X. Moreover, we also investigate effects of the number of hotspots and hot-times, and the effect of length of time window used to obtain them. Although the framework of our experiment can be regarded as an interest prediction method, we lay our emphasis on the performance of various feature sets under standard classification algorithms.

Usage detail records (UDRs) used in this paper span over 23 days, covering nine municipalities in the south of China. Each piece of UDR is generated when user accesses to mobile Internet via applications on his/her smartphone. The key fields and examples in our UDRs are provided in Table 2.2. Note that location consists of LAC and CELLID of a cellular tower where phoneNUM is served, and the corresponding longitude and latitude can be referred by a known translation table. For privacy, all phone numbers are translated into hash codes before we can reach them.

Each record is a four elements tuple. Who/when/where/what indicating that user phoneNUM contributed (end-start) seconds at location for mobile online activity in URL. Those four elements are typical and can be found in various kinds of datasets, such as news [31], Twitter [32], and the posts in discussion forums [33]. Besides, an increasing number of evidences, such as [34, 35], have shown that users tend to spend more time and energy on screen based activities than physical activities. Thus UDRs contain meaningful messages on human behavior, and provide fine-grained description for users' mobility and interest behavior.

To obtain ground truth data used for future training and validation, several challenges need to be considered. For one thing, due to the screen limitation of mobile devices, it is common to have mistaken operations in mobile internet usages. Thus not only the meaningful online behavior, but massive noises are also captured. For the other thing, the bipartite matrix user $\tilde{U}$RL is dramatically sparse due to the uniqueness of URL. Therefore this paper extracts the main website in URL to represent user interest, such as extracting cmread.com from wap.cmread.com in Table 2.2.

To obtain reliable and representative candidate users, we discard the individuals with less than 15 records everyday on average. When filtering candidate websites, duration time, frequency, and the number of coverage users are referred. Specifically, in every aspect, websites are ranked in descending order according to their values, and we select the subsets when the energy exceeds defined threshold $k \in [0, 1]$. And then we choose the websites if they exist in all the subsets generating from each aspect. Finally, we obtain more than 7 million records covering 179 candidate websites and 9720 candidate users in our valid dataset when we set $k = 0.9$.

Euclidean distance is not sensitive if variates vary in small intervals. Moreover, scale transformation methods seem hard to guarantee the fairness among all variates. To avoid these defects, we choose classification algorithms in [36] using entropy as an index in the process of modeling, namely DecisionTree (DT) and RandomForest (RF). In our cases, Gini impurity is used for both DT and RF in measuring the quality of a split. RF has 10 trees, and each of it with sqrt(nf) features (nf is the number of features that RF receives).

This chapter then utilizes precision, recall, and f1_score to measure the performance of different feature sets, which are defined in Eqs. (2.36)–(2.38).

$$precise = \sum_{l \in L} \omega_l p_l \tag{2.36}$$

$$recall = \sum_{l \in L} \omega_l r_l \tag{2.37}$$

$$f1 - score = \sum_{l \in L} 2\omega_l p_l r_l / (p_l + r_l) \tag{2.38}$$

Specifically, $p_l = TP_l / ((TP_l + FP_l))$ is the precision of activity l, and $r_l = TP_l / ((TP_l + FN_l))$ is the recall of activity l. L is the total candidate websites set; TP_l is the number of true positives of activity l; and FP_l is the number of false positives of activity l; FN_l is the number of false negatives of activity l. Finally, ω_l is the weight of activity l in the test dataset. This paper utilizes cross validation in our experiment to avoid overfitting. In particular, for each feature set, we split total data into K folders, in which K-1 folders are used for model training and the remains are used for validation. This process is repeated for K times so that each sample in the dataset is used both in model training and validation. We set K=10 in our experiment.

Next, we compare the performances of TCB among different feature sets $O/T/S$. Since TCB generates HS and HT, and both HS and HT contain corresponding statistical summaries, we also investigate the performance of sub-feature sets (namely HS and HT) generated by TCB and the effect of statistical summaries related to them. Without losing of generality, the number of hotspots m and hot-times n are set to 3, and the statistical window win is set to 7.

First of all, we investigate the performance of independent feature set, and the results are given in Fig. 2.20. Note that $X_{/\mathrm{STA}}$ indicates the feature set related X without corresponding statistical summaries. Regardless of metrics and classification algorithms, the tendencies of different feature sets are nearly identical in the process of cross validation, indicating that the feature sets used in DT and RF are stable and reliable. Compared to other feature sets, HS and HT rank the best, followed by $HS_{/\mathrm{STA}}$ and $HT_{/\mathrm{STA}}$, indicating that (1) feature sets generated by TCB are much informative and suitable for predicting mobile user interest, and (2) statistical information related to hotspots (hot-times) in HS (HT) is important for HS (HT) to achieve a better performance. On the contrary, the performances of traditional spatial–temporal feature sets S and T are even worse than that of original features O, implying that spatial–temporal features in single dimensionality are insufficient in predicting mobile user interest. Both spatial and temporal information should be taken into account. Moreover, the performances of O, S, T, HS, and HT are mainly consistent with the relation depicting in Fig. 2.20, indicating that the distance correlations between different feature sets and user interest are reliable.

Second, we examine the performance of dual combination of different feature sets. Note that values in Table 2.3 are the average after 10 runs in cross validation.

The results show that the performance of DT and RF is very similar. Compared to original feature set O (see Fig. 2.21), both S and T provide additional meaningful information for classification. However, the improvement seems to be limited. Besides, both the performance of OS and OT surpass the performance of ST, which means traditional spatial–temporal feature sets are redundant to each other, even less informative compared to original information recorded in O. On the contrary, feature sets generated by TCB are much more abundant, compared to the original and traditional spatial–temporal feature sets, and bring universal and remarkable improvement in dual combination cases. In particular, on temporality, integrating original information O, the precision improvement brought by HT is 38.3% compared to feature set T in the best performance (when RF is executed). While on spatiality, HS improves the precision by 36.1% compared to feature set S in the best performance. Moreover, although statistical summaries about hotspots and hot-time are meaningful in the prediction of user interest, original feature set O and traditional spatial feature set S can still make considerable compensation when statistical summaries are missing, while the traditional temporal feature set T is helpless. Finally, despite their impressive performance of HS and HT when

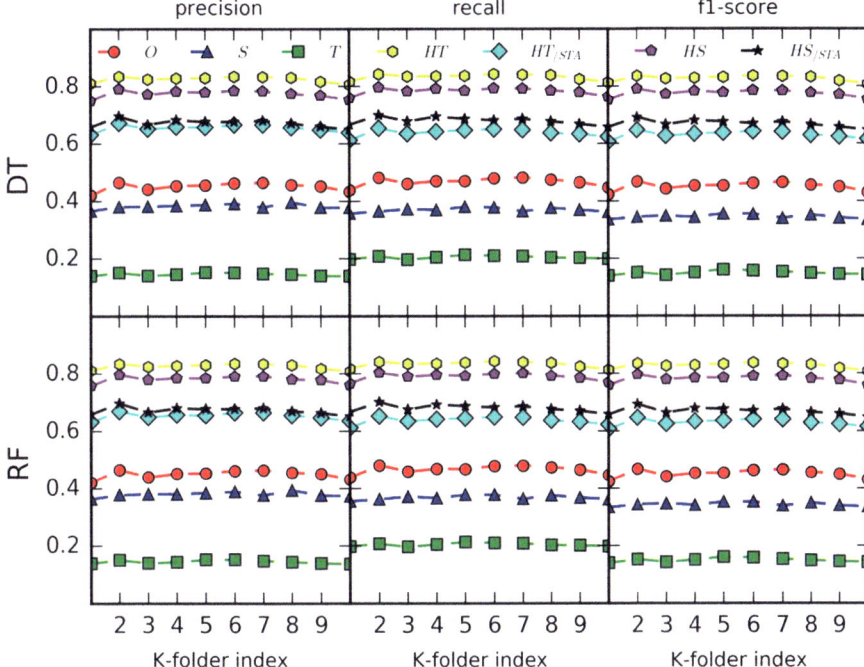

Fig. 2.20 The performance (precision/recall/f1-score) of DT and RF when different feature set is used

integrating O, S, and T, respectively, the combination of HS and HT doesn't show great improvement. An intuitive interpretation is that HS and HT are highly coupling in space–time. Since user mobility is constrained in a small area, mobile network usages occurred in hot-time are of great probabilities in hotspots.

Next, we investigate the performance of multi-combined feature sets, and how useful each feature set is in different combination cases. RF is used in modeling for simplicity. Original feature set O is used as starting line since it is the basic information in raw data. We also compare the performance of preference select proposed in [6]. For the simplicity, the probability of inertia is set to 0. User prefers to return to a historically interest category by the probability P_{ret}, and explore a new interest category with the probability $1-P_{ret}$. The historical interest category i is chosen according to the probability $\Pi_i = f_i$, in which f_i is the probability that how often interest category i is visited before. Then we vary $P_{ret} \in [0, 1]$ with interval 0.1, and present the best performance of preference selection. The results are shown in Fig. 2.7. All values produced by different feature sets are the average after 10 runs.

As it is shown in Fig. 2.21, HS and HT related feature sets bring universal improvement to user interest prediction. Compared to OST, OHSHT promotes the performance by 16.2%, and the figure is 13% when OHS/STAHT/STA is

Table 2.3 The performance (precision/recall/f1-score) of dual combination of different feature sets

		S	T	HT	HT/STA	HS	HS/STA
DT	O	0.436/0.431/0.432	0.45/0.45/0.449	0.809/0.814/0.81	0.756/0.763/0.758	0.796/0.801/0.797	0.758/0.763/0.759
	S		0.412/0.417/0.41	0.823/0.831/0.825	0.774/0.783/0.776	0.778/0.787/0.779	0.753/0.763/0.754
	T			0.818/0.815/0.816	0.549/0.544/0.542	0.771/0.768/0.768	0.644/0.645/0.643
	HT				0.825/0.833/0.826	0.808/0.814/0.809	0.805/0.811/0.806
	HT/STA					0.8/0.805/0.801	0.764/0.77/0.765
	HS						0.773/0.783/0.774
RF	O	0.448/0.456/0.442	0.434/0.444/0.435	0.817/0.823/0.819	0.77/0.778/0.772	0.809/0.814/0.81	0.763/0.77/0.765
	S		0.373/0.383/0.373	0.825/0.832/0.826	0.773/0.782/0.774	0.783/0.793/0.784	0.751/0.762/0.753
	T			0.831/0.834/0.831	0.54/0.537/0.528	0.785/0.789/0.785	0.631/0.635/0.629
	HT				0.826/0.833/0.827	0.815/0.819/0.815	0.813/0.818/0.814
	HT/STA					0.809/0.815/0.811	0.77/0.777/0.771
	HS						0.776/0.786/0.777

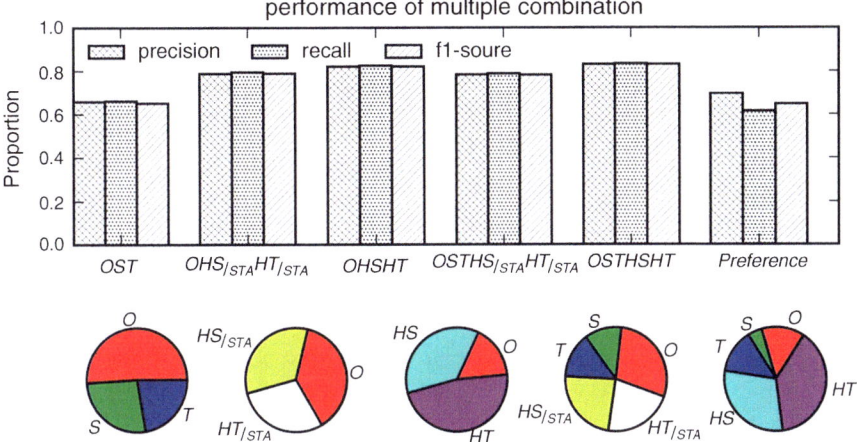

Fig. 2.21 Prediction performance comparison. The upper panel shows the performance of different feature sets. The lower panel shows the importance of different feature set in model fitting in corresponding classification task (from left to right)

used. By integrating HS and HT, the final precision of OSTHSHT can even reach 83%, generating 17.2% improvement compared to using OST alone. Without statistical summaries related to HS/HT, the improvement decreases to 12.6%, however, it is still remarkable. Note that the performance of $OHS_{/STA}HT_{/STA}$ and $OSTHS_{/STA}HT_{/STA}$ is very close, which indicates that traditional spatial–temporal features in S and T are redundant when taking $HS_{/STA}$ and $HT_{/STA}$ into account. Preference selection is only superior to OST, far less impressive than that combines HS and HT. We than go further and investigate how much different feature sets matter when multiple sets are used in model training from the perspective of feature importance [29]. The importance of each feature set is the sum of the corresponding features. Compared to traditional spatial–temporal feature sets S and T, HS and HT are more valuable in the process of modeling, which is obvious in the cases of OST and OHSHT. In the case of $OSTHS_{/STA}HT_{/STA}$, although the most important feature set is O, the performance improvement brought by the combination of O, S, T, $HS_{/STA}$, and $HT_{/STA}$ is about 30% compared to using O alone. Finally, in the case of OSTHSHT, HS and HT are the most significant feature sets in the process of classification modeling. The different importance distribution in $OSTHS_{/STA}HT_{/STA}$ and OSTHSHT also indicates that statistical information at hotspots and hot-times can make impressive contribution to users' interests prediction.

Since the number of hotspots/hot-times affects the collection of the effects received from hotspots/hot-times (HS/STA/ HT/STA), and the corresponding statistical summaries at hotspots/hot-times, in this section, we investigate the performance of HT/HT/STA/HS/HS/STA as a function of the number of hotspots/hot-times. For the simplicity of illustration, we refer hotspot and hot-time as centrality, and set win to 7. The results are shown in Fig. 2.22.

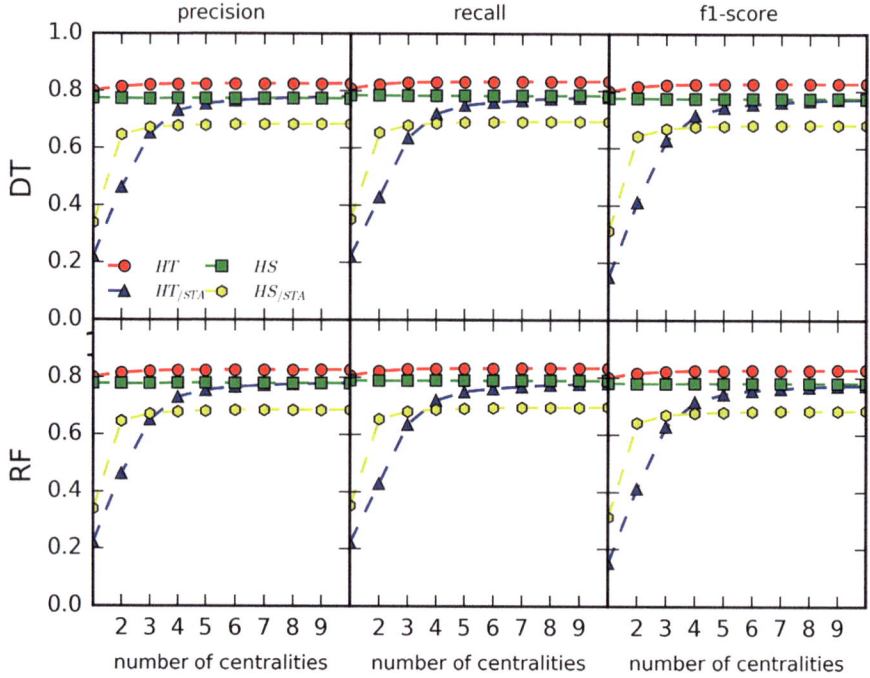

Fig. 2.22 Performance of HS/HS$_{\text{STA}}$/HT/HT$_{\text{STA}}$ when the number of hotspots/hot-times changes

At the very beginning, the performances of HS/STA and HT/STA show significant improvement when the number of centralities increases, and they reach saturation when the number of centrality gets larger. The saturation states for both feature sets show great advantages over $O/S/T$ (see Fig. 2.20). However, the variation of the number of centralities brings little effect when the corresponding statistical summaries are taken into account (see the performance of HS and HT), indicating that statistical summaries at centralities can make effective compensation when the number of centralities is limited. Moreover, the performance of HS is always superior to that of HS/STA, which implies statistical summaries at hotspots can't be replaced by a larger number of hotspots in HS$_{\text{STA}}$. This phenomenon also verifies that statistical summaries related to hotspots and hot-times can make significant contribution to the prediction of mobile user interests. To sum up, although a larger number of centralities can lead to a better prediction performance, it also means a higher dimensionality for data processing. With the help of statistical summaries at centralities, TCB can achieve relatively high performance even in a limited dimensionality, which makes it favorable in the era of mobile big data.

Since the hotspots/hot-times information and the corresponding statistical summaries are extracted in a time window win, we then investigate the performance of HT/HT$_{\text{STA}}$/HS/HS$_{\text{STA}}$ when time window win changes. A higher win means a larger amount of data needed for the process of extracting hotspots/hot-time related information. Without losing of generality, the granularity is set to day, and the number of hotspots m and hot-times n is set to 3. The results are shown in Fig. 2.23.

The results show that time widow has little effect on the performance of HT, indicating that few records are required to obtain hot-time information that used for projecting original temporal features into new vector space. On the other hand, the performance of HS, HS_{STA}, and HT_{STA} increases along with win. Although a larger win indicates a detailed description for mobile network usage, the improvements are not significant when win is larger than 7. In conclusion, TCB is effective and only requires a small amount of data for building feature transformation space, which makes it quite applicable in the scenarios of mobile big data.

2.2.2   Context-Aware Feature Discovery

2.2.2.1   Design of UIAHW

We will introduce design of UIAHW from three aspects, including design, challenges, and solutions.

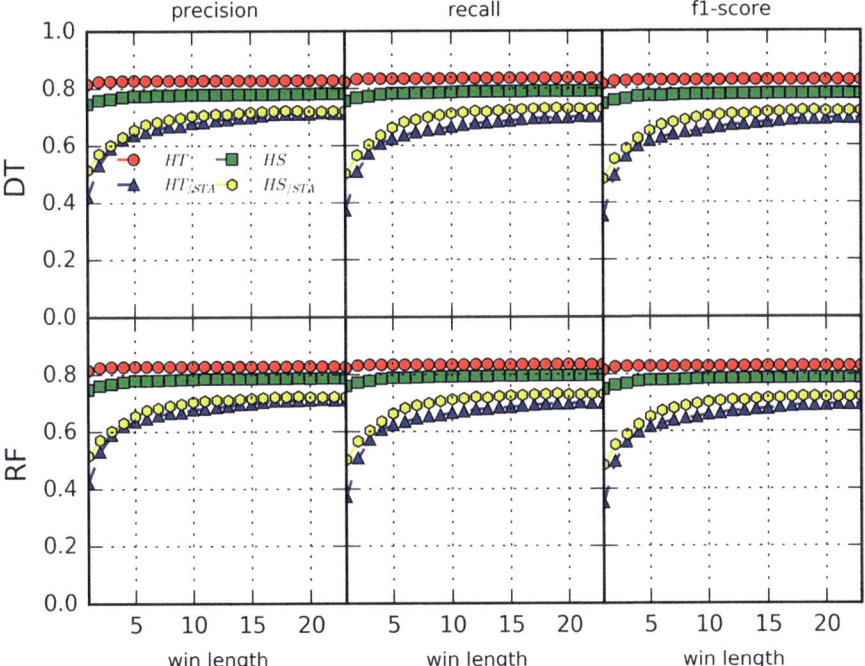

Fig. 2.23 Performance variation when win length changes

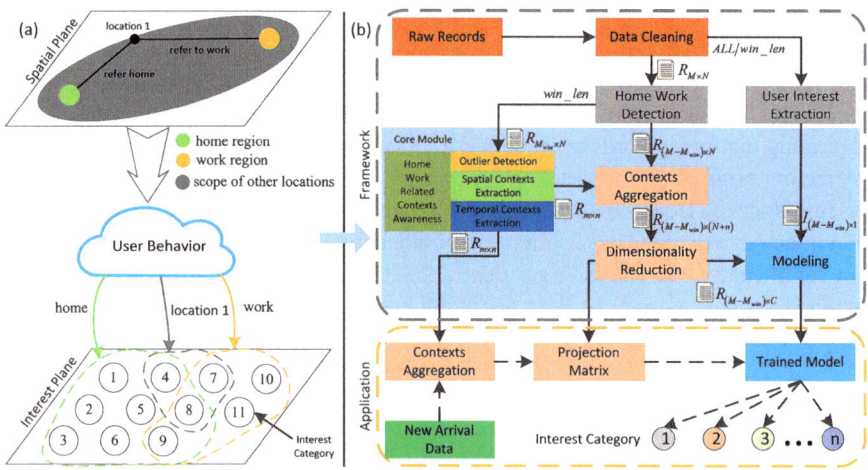

Fig. 2.24 (**a**) A typical case using home and work places to identify user interests at location 1. (**b**) The framework and application of UIAHW. The procedure of data cleaning filter records with key values missing. R shows the change of dimension in different steps. M is the total number of records. N is the number of features. M_{win} the number of records in time window win_{len}. m is the number of users. n is the number of home–work related contexts. C is the number of main components and $C \ll N + n$. The dimension of projection matrix is $(N + n) \times C$

UIAHW consists of home–work detection, user interest extraction, and the core module, which includes home–work related contexts awareness, contexts aggregation, dimensionality reduction, and modeling. The detailed architecture and data flow of UIAHW are shown in Fig. 2.24b. Generally, UIAHW should have the following capacities:

- Universality: UIAHW only requires the information concerning who, when, and where, and they are quite easy to be acquired without further process. The logic position of UIAHW is located at the physical layer, which means no specific software should be adapted for preparing the input data.
- Flexibility: The components of UIAHW should be replaceable and independent of the structure and type of dataset.
- Easy deployment: UIAHW has no harsh requirements for data storage and computing power when the principle components and the projection matrix are learnt. Actually, the complexity of obtaining principle components and projection matrix is up to the algorithms used in the process of dimensionality reduction.

Specifically, in order to acquire user interests, the components of UIAHW should accomplish the following tasks:

- Preprocessing: Filter broken records with important value; detect home and work place for mobile users; and prepare interests sample dataset used both for modeling and validation.

- Contexts extension and reduction: Based on the anchors of home and work locations, retrieve the network usage instance and obtain as many contexts as possible to describe user behavior, select the most principle information for future modeling, and generate projection matrix.
- Mapping: Connect user interests with selected principle information using state-of-the-art machine learning algorithms.
- Application: Based on the obtained home–work related contexts, projection matrix, and trained model, new data should be transformed onto new subspace which is dominated by principle information. And user interests are acquired by the trained model.

The following challenges in UIAHW are still open issues and attract extensive attentions [37–41]. For the simplicity of illustration, we utilize general and replaceable method for preparing the required data for the core module of UIAHW:

- Important location identification

  Different from the work in [40, 41], we design our home–work detection algorithms similar to the theory of squeeze theorem. For every user, the longitude and latitude of first and last record of every day are collected and put into candidate_1 and candidate_2. The most frequent record in candidate_1 and candidate_2 is chosen as home_1 and home_2, respectively. Then the home of current user is located in the middle of home_1 and home_2. As for work location detection, we first split work time into morn and noon. Similarity, for each user, the most frequent location in morn and noon in every day is obtained and put into candidate_1 and candidate_2, respectively. Then the most frequent record in candidate_1 and candidate_2 is collected as work_1 and work_2. The work of current user is located in the middle of work_1 and work_2.

  We then refer to the tie strength theory proposed in [11] to investigate network usage. Let $S_i = s_i^1, s_i^2, \ldots, s_i^n$ denote the locations that user i visited. Then the strength of tie between user i and base station j is $t_{ij} = nT_{ij}/(\sum_{k=1}^{n} T_{ik})$, in which T_{ik} is the total time that user i contributes to base station k. We then define the strong tie when $t_{ij} \geq 1$, and weak tie when $0 < t_{ij} < 1$. We plot the complementary cumulative distribution function (CCDF) of different types of tie in terms of the distance from home/work, as shown in Fig. 2.25a, b. From the perspective of heavy network demand, the results show that home > work > other locations. This phenomenon also inspires us that home–work related contexts may provide much more information than record contexts in identifying user interests.
- User Interest Extraction

  A fitting power law appears when we plot the time received by websites in descending order (see Fig. 2.25c), which means only few websites are well known to most of the users and dominant in their mobile internet usage. We define these websites as candidate websites, and they are ranking in the top $k\%$ from the perspective of both the length of receiving time and the number of coverage users. A larger k means that candidate websites are closer to real situation; however, the size of candidate websites increases dramatically. So

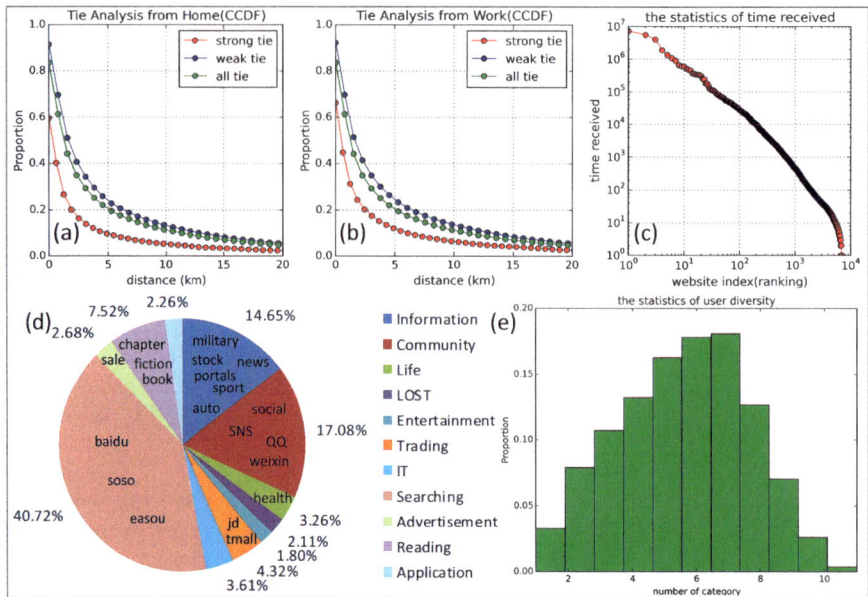

Fig. 2.25 The statistics of user network usage (**a, b**) and interest (**c–e**). (**a**) and (**b**) Show tie distribution from home and work places. (**c**) Is the distribution of time received for all websites in loglog coordinate. (**d**) denotes ratio of each category from the perspective of receiving time, some typical labels of main categories are embedded in the corresponding areas. Websites that can't be visited are grouped into a category named LOST since they indicate abnormal. (**e**) Shows the distribution of user interest diversity. Users with extreme interest diversity are rare, and most users have 6 or 7 interests

we set $k = 6$ to make a trade-off between reality and complexity. Finally, 271 websites are selected as candidate websites, and 20 qualified volunteers are recruited to manually classify candidate websites. Then labels are clustered into category to reduce concept overlapping. The results are shown in Fig. 2.25d, e.

2.2.2.2   Connecting User Interest by Adding Home and Work Related Contexts

The general record contexts only contain information concerning who, when, and where. Since home and work regions attract most of users' mobile network usage and occupy most of their daily lives, we can have a better understanding of user behavior from temporal, spatial, and interest aspects. In fact, when we employ UIAHW without using the process of dimensionality reduction on our test data in python environment using scikit-learn [36], the feature importance index of home and work related contexts accounts for 59%, more than 18% compared to the result of record contexts.

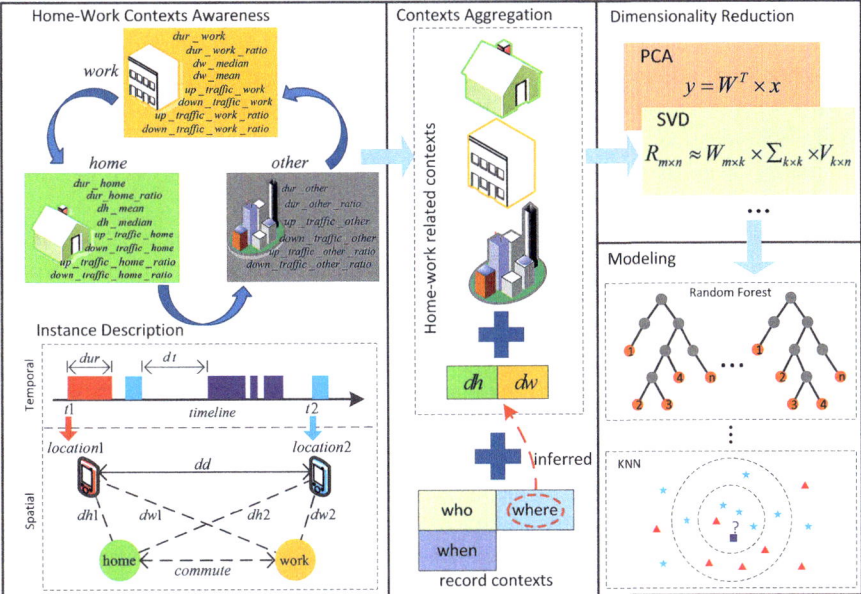

Fig. 2.26 The core models in UIAHW. Instance description shows a typical mobile Internet usage. Contexts aggregation combines both record contexts and home–work related contexts as an entire one. Various classical methods can be used in the step of dimensionality reduction and modeling, which makes UIAHW very flexible at different scenarios

As shown in Fig. 2.26, according to data processing flow, based on home and work location, home–work related contexts are mined in the process of home and work contexts awareness. In contexts aggregation procedure, each record can be significantly extended by home and work related contexts generated by the corresponding user. Since the aggregated contexts may introduce redundant information, dimension reduction is applied both for computation and store complexity. Then the filtered principal components are used for modeling.

This chapter separates the spatial plane into three parts, namely home, work, and other locations. Contexts concerning those three parts are home and work related since the sum of the corresponding context values in home, work, and other locations is unit value. Some of the typical open issues are how many contexts have strong correlation with user interests, and what are they. However, those answers are still mysterious and hidden in user behaviors. So we extract home and work related contexts using greedy strategy by reconstructing user network usage.

A typical mobile Interest usage behavior can be presented in instance description in Fig. 2.26. For a given user Jim, his commute distance is known when his home and work location is obtained. Assume that he accessed mobile Internet at *t1* and *t2*. At the moment *t1*, Jim's interest is category 1 (shown in the red color) and lasts for seconds. He used mobile Internet at location1 and it has certain longitude and latitude coordinates. However, we can also infer the current mobile scope by referring to home and work places. Let location1 be *dh1* and *dw1* kilometers away from his home and work place, respectively. up_traffic_1 and down_ traffic_1 are corresponding traffic consumption in uplink and downlink. At the moment *t2*, his interest turn to category 2 (shown in the blue color) and he moved to location2. Similarity, location2 also reflects Jim's mobile scope, and it is *dh2* and *dw2* kilometers away from his home and work place. The distance between location1 and location2 is *dd*.

We collect all relevant contexts mentioned above in a time window denoted by win_len (unit: day), and generate three groups of contexts based on home, work, and other locations for each user.

At this step, we extend records by attaching home and work related contexts generated by same user, and append the mobile scope *dh*, according to the longitude and latitude in original record and the referred home–work places. The philosophy behind this procedure is that: for different users, home and work related contexts can characterize mobile users from the aspects from temporal, spatial, and traffic consumption; for the same user, the usage behavior recorded in the original records may have certain relation with his interest, as it investigated in [41]. However, results in Fig. 2.27 show that home–work related contexts make user interest identification more predictable.

The dimension reduction is applied right after the contexts aggregation for the two reasons: the process of home and work related contexts awareness employs

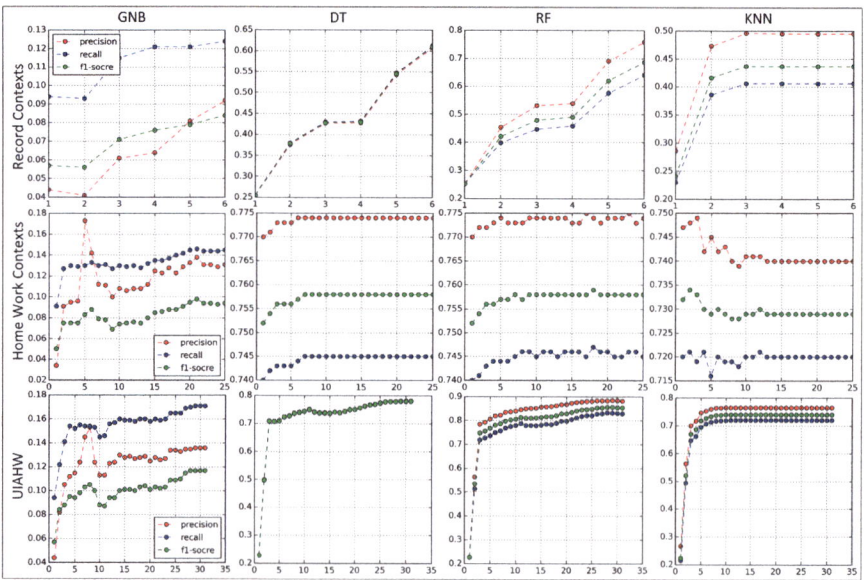

Fig. 2.27 The performances of UIAHW when typical classification algorithms are used for different contexts. The horizontal axis in each figure denotes the change of principle components number left after dimensionality reduction

greedy strategy and may introduce redundant information; the dimension gets much larger than that in original records. Both situations will bring huge burden into computation and data storage. There are many well-known algorithms for dimensionality reduction, such as PCA, kernel PCA, SVD, and tensor analysis [42]. This article selects PCA to make a trade-off between accuracy and complexity. The procedure of dimensionality reduction not only reduces the data dimension without losing much information, but also obtains a projection matrix W to transform original data onto a new subspace, which is useful for future application.

User interests and the principle components of user contexts (including record and home–work related contexts) are connected in the subspace dominated by principle components. Various state-of-the-art machine learning algorithms can be used in this procedure. In particular, we utilize the most popular classification algorithms concluded in [43], such as Naïve Bayes, CART, and KNN, and compare the performance of UIAHW when different algorithms used in the process of modeling. Besides, RandomForest [44] is also selected due to its remarkable performance. In various scenarios, the criterion for choosing applicable algorithm is to weigh accuracy against complexity.

2.2.2.3 Evaluation

UIAHW is validated on usage detail records (UDRs) dataset with more than 3.2 million records, covering 6800 residents of a developed city in China spanning 23 days. Since scikit-learn contains various state-of-the-art machine learning algorithms [36], so UIAHW runs on Python environment using scikit-learn module. We evaluate the performance of UIAHW from the perspective of effectiveness and metrics, which includes precision, recall, and f1-score.

Precision, recall, and f1-score are selected as metrics to evaluate the performance of UIAHW when typical algorithms, namely GaussianNB(GNB), Decision-Tree(DT), RandomForest (RF), and K-Nearest Neighbors(KNN), are used in the modeling process. All metrics range from 0 to 1, and a higher value means a better performance. The number of trees in RF is set to 10 to avoid huge time and memory consumption. The *win_len* is set to 7. The results are presented in Fig. 2.27.

Compared to using record contexts alone, home–work related contexts contain much more information to identify user interests. Moreover, the performances of UIAHW rank the best. UIAHW brings universal performance improvement in spite of different algorithms used in the procedure of modeling. In particular, RF shows the best performance, and the final precision can reach 88.75%. However, it is not cost-effective when the number of components is larger than 7 due to the complexity (Fig. 2.28).

To investigate the effectiveness of UIAHW, we split the whole dataset into training and test sets and analyze the performance of UIAHW over win_len. A higher win_len means a larger amount of data needed for the process of home–work related contexts extraction.

Fig. 2.28 The effectiveness of UIAHW. The changes from 1 to 23 (the whole period of observation). RF is selected in the step of modeling due to its best performance (see Fig. 2.27). The number of principle components is set to 7 for the trade-off between performance and complexity

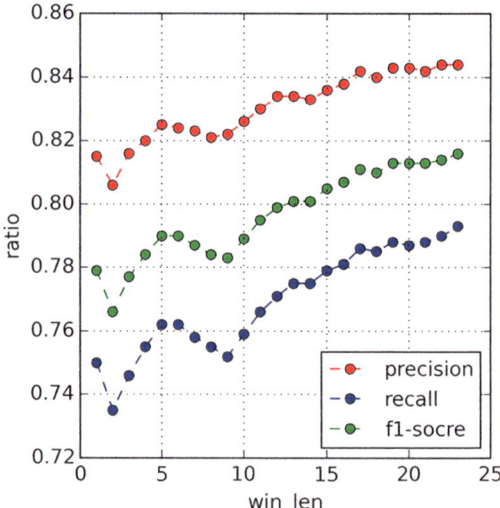

The performance of UIAHW fluctuates and grows little when win_len increases. However, even the precision in the worst case is still more than 0.8. It indicates that a small amount of data used for home–work related contexts extraction can still achieve a relatively impressive performance. Thus UIAHW is effective and has no harsh requirements for data storage.

References

1. Burke, J., Estrin, D., Hansen, M., Parker, A.: Participatory Sensing. 1st Workshop on Wireless Sensor Web (WSW06), 1–5 (2016)
2. Estrin, D.: Participatory sensing: applications and architectures. In: 8th International Conference on Mobile Systems, Applications and Services (MobiSys10), pp. 3–4. IEEE, Piscataway (2010)
3. Sassi, A., Zambonelli, F.: Coordination Infrastructures for future smart social mobility services. IEEE Intell. Syst. **29**(5), 78–82 (2014)
4. Hsu, W.-J., Dutta, D., Helmy, A.: CSI: a paradigm for behavior-oriented profile-cast services in mobile networks. Ad Hoc Netw. J. **10**(8), 1–14 (2012)
5. Cover, T.M., Thomas, A.J.: Elements of Information Theory. Wiley, New York (1991)
6. Zheng, Y., Xie, X., Ma, W.-Y.: GeoLife: a collaborative social networking service among user, location and trajectory. IEEE Data Eng. Bull. **33**(2), 32–40 (2010)
7. Reddy, S., Estrin, D., Srivastava, M.: Recruitment framework for participatory sensing data collections. Pervasive Comput. Lect. Notes Comput. Sci. **6030**, 138–155 (2010)
8. Viswanath, P.: Opportunistic Communication: A System View. Department of Electrical and Computer Engineering. University of Illinois, Illinois (2014)
9. Liang, Q., Cheng, X., Huang, S.C.H., Chen, D.: Opportunistic sensing in wireless sensor networks: theory and application. IEEE Trans. Commun. **63**(8), 2002–2010 (2014)
10. Blum, J., Eskandarian, A., Hoffman, L.: Challenges of intervehicle ad hoc networks. IEEE Trans. Intell. Transp. Syst. **5**, 347–351 (2004)

11. Grossglauser, M., Tse, D.: Mobility increases the capacity of ad hoc wireless networks. IEEE/ACM Trans. Netw. **10**(4), 477–486 (2002)
12. Fall, K.: A delay-tolerant network architecture for challenged internets. ACM SIGCOMM 27–34 (2003)
13. Fletcher, R., Hasson, A.: DakNet: rethinking connectivity in developing nations. IEEE Comput. **37**(1), 78–83 (2004)
14. Hull, B., Bychkovsky, V., Zhang, Y., Chen, K., Goraczko, M., Miu, A., et al.: Cartel: a distributed mobile sensor computing platform. ACM SenSys 125–138 (2006)
15. DTN Research Group (DTNRG). http://www.dtnrg.org/
16. Sutton, R.S., Barto, A.G.: Reinforcement Learning: An Introduction. MIT Press, Cambridge (1998)
17. Cosimo, P., Tuzhilin, A., Gorgoglione, M.: Using context to improve predictive modeling of customers in personalization applications. IEEE Trans. Knowl. Data Eng. **20**(11), 1535–1549 (2008)
18. Feng, W., Elmirghani, J.M.H.: Green ICT: energy efficiency in a motorway model. In: 3rd International Conference on Next Generation Mobile Applications, Services, and Technologies, pp. 389–394. IEEE, Piscataway (2009)
19. Mitchener, W., Vahat, A.: Epidemic Routing for Partially Connected Ad Hoc Networks. Technical Report CS-2000-06. Duke University, Durham (2000)
20. Jain, S., Fall, K., Patra, R.: Routing in a delay tolerant network. ACM SIGCOMM 145–157 (2004)
21. Van Mieghem, P., Blenn, N., Doerr, C.: Lognormal distribution in the digg online social network. Eur. Phys. J. B **83**(2), 251–261 (2011)
22. Anirban, M., et al.: A tale of the tails: power-laws in internet measurements. Network IEEE **27**(1), 59–64 (2013)
23. Chaoming, S., et al.: Limits of predictability in human mobility. Science **327**(5968), 1018–1021 (2011)
24. Granovetter, M.S.: The strength of weak ties. Am. J. Sociol. 1360–1380 (1973)
25. Wenjun, W., et al.: A comparative analysis of intra-city human mobility by taxi. Physica A: Statistical Mechanics and its Applications. 420, 134–147(2015)
26. Zhao, Z.-D., Zhou, T.: Empirical analysis of online human dynamics. Phys. A Stat. Mech. Appl. **391**(11), 3308–3315 (2012)
27. Sibren, I., et al.: Identifying Important Places in People's Lives from Cellular Network Data. Pervasive Computing, pp. 133–151. Springer, Berlin (2011)
28. Szkely, G.J., Rizzo, M.L., Bakirov, N.K.: Measuring and testing dependence by correlation of distances. Ann. Stat. **35**(6), 2769–2794 (2007)
29. Li, R., Zhong, W., Zhu, L.: Feature screening via distance correlation learning. J. Am. Stat. Assoc. **107**(499), 1129–1139 (2012)
30. Rizzo, M.L., Szekely, G.J.: Energy: E-statistics (energy statistics). R package version 1.6.2(2014). http://CRAN.R-project.org/package=energy
31. Gabor, S., Huberman, B.A.: Predicting the popularity of online content. Commun. ACM. **53**(8), 80–88 (2010)
32. Asur, S., Huberman, B.A., Szabo, G., Wang, C.: Trends in social media: Persistence and decay. In: Proc. of the 5th International Conference on Weblogs and Social Media (ICWSM), p. 434. AAAI Publications, Palo Alto (2011)
33. Christina, A., Huberman, B.A., Wu, F.: Harvesting collective intelligence: temporal behavior in yahoo answers. arXiv preprint arXiv: 1001.2320 (2010)
34. Serrano-Sanchez, J.A., et al.: Associations between screen time and physical activity among Spanish adolescents. PloS One **6**(9), e24453 (2011)
35. Edward, A., et al.: 45-Year trends in women's use of time and household management energy expenditure. PloS One **8**(2), e56620 (2013)
36. Pedregosa, et al.: Scikit-learn: machine learning in python. JMLR **12**, 2825–2830 (2011)
37. Bing, L., Lam, W., Wong, T.L., Jameel, S.: Web query reformulation via joint modeling of latent topic dependency and term context. ACM Trans. Inf. Syst. (TOIS) **33**(2), 6 (2015)

38. Wu, X., Zhu, X., Wu, G.Q., Ding, W.: Data mining with big data. IEEE Trans. Knowl. Data Eng. **26**(1), 97–107 (2014)
39. Shafiq, O., Alhajj, R., Rokne, J.G.: On personalizing web search using social network analysis. Inf. Sci. **314**, 55–76 (2015)
40. Yuan, Q., Cong, G., Zhao, K., Ma, Z., Sun, A.: Who, where, when, and what: a nonparametric Bayesian approach to context-aware recommendation and search for twitter users. ACM Trans. Inf. Syst. (TOIS) **33**(1), 2 (2015)
41. Trestian, I., Ranjan, S., Kuzmanovic, A., Nucci, A.: Measuring serendipity: connecting people, locations and interests in a mobile 3G network. In: Proceedings of the 9th ACM SIGCOMM conference on Internet measurement conference. pp.267–279. ACM, New York (2009)
42. Zhang, Y., Chen, M., Mao, S., Hu, L., Leung, V.: Cap: community activity prediction based on big data analysis. Netw. IEEE. **28**(4), 52–57 (2014)
43. Xindong, W., et al.: Top 10 algorithms in data mining. Knowl. Inf. Syst. **14**(1), 1–37 (2008)
44. Leo, B.: Random forests. Mach. Learn. **45**(1), 5–32 (2001)

Chapter 3
Mobile Data Application in Wireless Communication

Abstract In recent years, the cellular data traffic has been dramatically increased due to the explosive growth in mobile applications, putting a lot of pressure on capacity improvement of cellular networks, and the limited licensed spectrum is still the principal bottleneck for capacity improvement. To tackle this problem, traffic loading from cellular to WiFi has been recently developed. However, the spectrum utilization of the WiFi network is bound to suffer from potential packet collisions due to its contention-based access protocol, especially when the number of competing WiFi users grows large. To tackle this problem, we propose to transfer some WiFi users to be served by the LTE system. This chapter will be elaborated in turn: mobile date offloading from cellular to WiFi and mobile date offloading from WiFi to cellular.

Keywords Cellular data traffic · Explosive growth · Spectrum · Traffic loading · WiFi · Packet collisions · LTE

3.1 Mobile Data Offloading in LTE and WiFi Systems

Traffic offloading is a kind of common method for delivering cellular data traffic over unlicensed bands. In this chapter, we first develop a hybrid method to take full advantages of both traffic offloading and resource sharing methods, where cellular base stations (BSs) offload traffic to WiFi networks and simultaneously occupy certain number of time slots on unlicensed bands. Then, we analytically compare the cellular throughput with the guarantee of WiFi per-user throughput in the single-BS scenario. We find that traffic offloading can achieve better performance than resource sharing when existing WiFi user number is below a threshold and the hybrid method achieves the same performance as the resource sharing method when existing WiFi user number is large enough. In the multi-BS scenario where the coverage of small cells and WiFi access points are mutually overlapped, we consider to maximize the minimum average per-user throughput of each small cell and derive a closed-form expression for the throughput upper bound in each method. Meanwhile, practical traffic offloading and resource sharing algorithms are also

© Springer Nature Switzerland AG 2019
H. Jiang et al., *Mobile Data Mining and Applications*, Information Fusion and Data Science, https://doi.org/10.1007/978-3-030-16503-1_3

developed for the three methods. Numerical results validate our theoretical analysis and demonstrate the effectiveness of the proposed algorithms as well.

3.1.1 Mobile Data Offloading: Concepts and Applications

The cellular data traffic has been dramatically increased in the past few years due to the explosive growth in mobile applications, putting a lot of pressure on capacity improvement of cellular networks. To deal with it, many new techniques have been introduced for long-term evolution (LTE) and LTE-advanced networks, such as massive multiple-input multiple-output (MIMO), heterogeneous networks with small cells, direct device-to-device communications, etc. Despite these cutting-edge techniques, the limited licensed spectrum is still the principal bottleneck for capacity improvement.

To tackle this problem, several methods have been recently developed to use the unlicensed bands for delivering cellular data traffic. Among them, traffic offloading is the most common one, which offloads the data initially targeted for cellular systems to WiFi networks. However, how to guarantee the quality of service (QoS) of cellular traffic is a challenging issue since WiFi is operated in unlicensed bands and difficult to provide QoS due to the distributed coordination function (DCF) protocol. Moreover, the amount of the offloaded traffic should be carefully designed to avoid oversaturation and excessive packet collisions in the WiFi network.

Mobile data offloading has been widely investigated in recent years [1]. A quantitative study on offloading cellular traffic to WiFi networks has been presented in [2]. The authors in [3] have developed a distributed cross-system learning framework to improve the cellular throughput by offloading delay-tolerant traffic to WiFi networks. Various WiFi offloading algorithms have been developed to improve the system performance [4–6]. Besides inter-network data offloading, intra-network data offloading has also been investigated where the macro base station (BS) offloads traffic to femtocells. A tractable model to analyze the effects of traffic offloading has been developed for heterogeneous networks in [7]. The load balancing through user association has been investigated [8]. Economic issues of data offloading in femtocells networks have been investigated in [9] and [10].

As shown in Fig. 3.1, in this chapter, we consider a cellular system with small cell base stations (SBSs) coexisting with WiFi networks. Each SBS is overlapped with several WiFi access points (APs) and each AP is also overlapped with several SBSs. The SBSs transmit on certain licensed bands. However, they can also offload traffic to the APs overlapped with them and/or share their unlicensed bands.

To avoid inter-system interference, orthogonal resource sharing is considered, which can be guaranteed by the LBT protocol or the duty-cycle method. Here, we employ the duty-cycle method, where some time slots on the unlicensed bands are occupied by the cellular system, while the others are used by WiFi [11, 12]. In such a coexisting network, we will address the following three problems: (1) Traffic offloading or resource sharing, which one is better? (2) How can we jointly consider

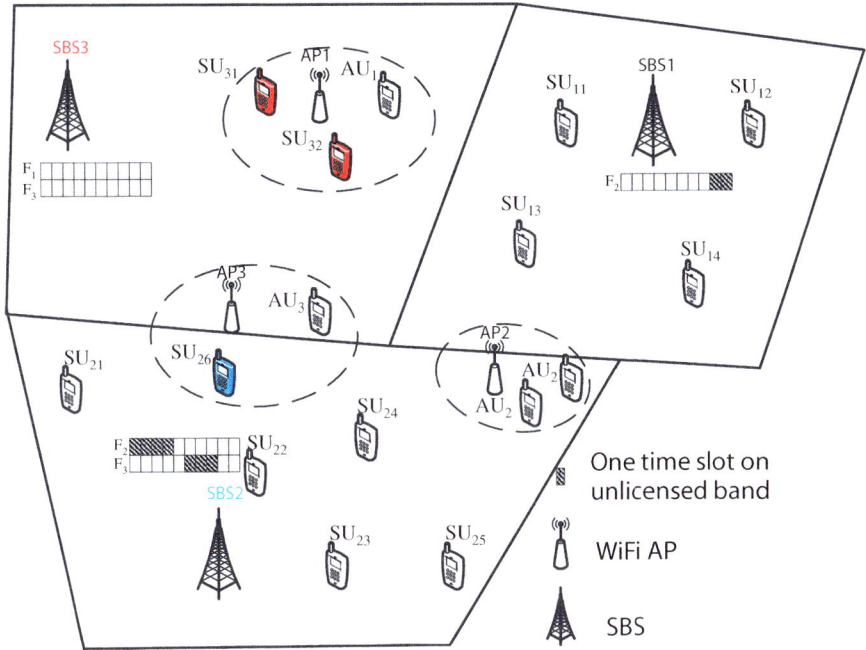

Fig. 3.1 System model for LTE and WiFi coexistence

both to further improve the performance? (3) How to design traffic offloading and resource sharing algorithms to maximize the LTE performance while guaranteeing the per-user throughput of WiFi networks?

The main contributions of the chapter include:

- To further improve the performance by taking full advantages of both traffic offloading and resource sharing methods, a hybrid method is developed where SBS can offload traffic to WiFi and simultaneously occupy certain number of time slots on unlicensed bands.
- We analytically compare the LTE throughput in the three methods, namely traffic offloading, resource sharing, and the hybrid method, in the single-BS scenario with the minimum WiFi per-user throughput guaranteed. It is proved that traffic offloading outperforms resource sharing when the existing WiFi user number is below a threshold and the hybrid method achieves the same performance as the resource sharing method when the existing WiFi user number is greater than a threshold.
- We analytically compare the LTE throughput in the three methods, namely traffic offloading, resource sharing, and the hybrid method, in the single-BS scenario with the minimum WiFi per-user throughput guaranteed. It is proved that traffic offloading outperforms resource sharing when the existing WiFi user number is below a threshold and the hybrid method achieves the same performance as the

resource sharing method when the existing WiFi user number is greater than a threshold.

The rest of this chapter is organized as follows: The system model and problem formulation are presented in Sect. 3.1.2. We will investigate the single-BS scenario and the multi-BS scenario in Sects. 3.1.3 and 3.1.4.

3.1.2 System Model and Problem Description

In this section, we first describe the LTE and WiFi coexistence system model and introduce the three different unlicensed spectrum utilization methods. After that, we will formulate the analytical problem for each method.

3.1.2.1 System Model

As illustrated in Fig. 3.1, we consider a scenario with M SBSs, $M = \{1, 2, \ldots, M\}$, and K APs, $K = \{1, 2, \ldots, K\}$. Both SBSs and APs are randomly located according to the *Poisson point process* (PPP) model. The coverage region of SBS can be modeled via a Voronoi tessellation. Denote a_{mk} as the overlapping indicator where $a_{mk} = 1$ indicates that SBS m is overlapped with AP k and $a_{mk} = 0$ otherwise. Let Φ_m denote the set of APs that have overlapping coverage with SBS m. We further assume that each SBS is allocated with B_m^S licensed bandwidth to serve N_m^S users and there are N_k^A users in AP k.

The WiFi network is assumed to support the IEEE 802.11n protocol, working in the 5 GHz band. There are several channels with a bandwidth of B and each AP can dynamically select an available channel with less interference to transmit. We further assume that the APs overlapping with the same SBS will choose different channels to avoid severe interference due to geographical proximity. This is feasible since the WiFi channels are generally sufficient to support different APs in one SBS, e.g., there are 23 channels for IEEE 802.11n in the 5 GHz band. As illustrated in Fig. 3.1, AP1, AP2, and AP3 will use unlicensed channel F1, F2, and F3, respectively.

The following three methods are investigated:

- Traffic offloading: In this method, SBS m will offload N_{mk} users to AP k if $k \in \Phi_m$. We assume that the maximum number of users that can be offloaded from SBS m to AP k is $N_{mk}^{\max}$, which can be determined by the number of those cellular users in SBS m that are also located within the coverage of AP k and is known a priori.
- Resource sharing: In this method, the unlicensed band is divided into a total of L^A time slots whose length is long enough to transmit one LTE frame. SBS m will occupy $\tilde{L}_{mk}$ of L^A time slots from AP k [13]. For convenience, we normalize

$L^{mk} = \frac{\tilde{L}_{mk}}{L^A}$ ($L_{mk} \in [0, 1]$) and assume that each SBS can coordinate such that the time slots occupied from different APs will not overlap.

- The hybrid method: In this method, SBS can both offload users to WiFi and at the same time occupy time slots on unlicensed bands. Specifically, SBS m will offload N_{mk} users to AP k and occupy its L_{mk} time slots.

As in Fig. 3.1, SBS3 employs the traffic offloading method, which offloads two cellular users, SU$_{24}$ and SU$_{32}$, to AP1, SBS1 utilizes the resource sharing method, occupying two time slots of AP2, and SBS2 uses the hybrid method, which offloads SU$_{26}$ to AP3 and at the same time occupies three time slots of AP3 and four time slots of AP2.

In this chapter, we assume that traffic offloading and resource sharing are coordinated by a central controller, which has all information, such as the locations and the number of users in each SBS. Moreover, the number of users in each AP can be estimated by the method proposed in [14]. It should be noted that the SBS can also share the unlicensed spectrum with those APs beyond its coverage. However, complicated resource scheduling among different SBSs is required. Therefore, we do not consider this situation in the chapter.

3.1.2.2 WiFi Throughput Analysis

The carrier sense multiple access with collision avoidance (CSMA/CA) scheme with binary slotted exponential backoff is adopted in WiFi networks. The saturation throughput of a WiFi network with n users represents the maximum load that the system can carry in stable conditions, which can be analyzed with a discrete time Markov chain (DTMC) model developed in [15]. Note that although we focus on the saturation system, our analysis can be also extended into the non-saturation system after some minor modifications since both systems have very similar WiFi throughput models [16].

Let P_{tr} and P_s be the probability that there is at least one transmission in a slot time and a transmission occurring on the channel is successful, respectively, which can be expressed as

$$P_{\text{tr}} = 1 - (1 - \tau)^n , \tag{3.1}$$

$$P_{\text{s}} = n\tau (1 - \tau)^{n-1} / P_{\text{tr}} \tag{3.2}$$

where τ is the transmission probability of each user. Then, from [15], the saturation throughput of the WiFi network can be expressed as

$$R(n) = \frac{P_{tr} P_s E[P]}{(1 - P_{tr}) T_\sigma + P_{tr} P_s T_s + P_{tr} (1 - P_s) T_c} \tag{3.3}$$

where T_s is the average time the channel is sensed busy because of a successful transmission, T_c is the average time the channel is sensed busy by each station during a collision, T_σ is the duration of an empty slot time, and $\mathbb{E}\,[P]$ is the average packet size.

To compute Eq. (3.3) for a given DCF access mechanism, it is necessary to specify the corresponding values T_c and T_s. There are two packet transmission schemes employed by DCF, namely the basic mechanism and the RTS/CTS access mechanism. T_c and T_s in these two mechanisms can be, respectively, expressed as [15]

$$
\begin{cases}
T_s^{\text{bas}} = (\text{H}+E\,[P])\,/C+\text{SIFS}+\delta+\text{ACK}/C+\text{DIFS}+\delta, \\
T_c^{\text{bas}} = \left(\text{H}+E\left[P^*\right]\right)/C+\text{DIFS}+\delta,
\end{cases}
$$

$$
\begin{cases}
T_s^{\text{rts}} = \text{RTS}/C+\text{SIFS}+\delta+\text{CTS}/C+\text{SIFS}+\delta+(\text{H} \\
\qquad\quad +E\,[P])/C+\text{SIFS}+\delta+\text{ACK}/C+\text{DIFS}+\delta, \\
T_c^{\text{rts}} = \text{RTS}/C+\text{DIFS}+\delta
\end{cases}
$$

where $\text{H} = \text{PHY}_{hdr} + \text{MAC}_{hdr}$, $E\left[P^*\right]$ is the average length of the longest packet traffic involved in a collision, C is the WiFi channel bit rate, and ACK, DIFS, δ, RTS, and CTS are WiFi parameters. Although IEEE 802.11n supports various channel data rates, we could assume that C is constant in the above equations since the channel data rate is a system-level parameter and does not frequently change with user distribution, traffic pattern, and channel status. In other words, the time-scale of changing the data rate is much longer than that of executing the traffic offloading or resource sharing algorithm.

Note that the expression in Eq. (3.3) ignores the hidden terminal. However, our results can be easily extended into that scenario with hidden terminals since WiFi with hidden terminals has a similar throughput expression as Eq. (3.3) according to [17]. Furthermore, even if the expression in Eq. (3.3) is for the uplink, the developed algorithms can be also used in the downlink.

3.1.2.3   LTE Throughput Analysis

Both traffic offloading and resource sharing will definitely degrade the performance of the WiFi network. In this chapter, we aim to maximize the LTE throughput while guaranteeing the performance of the WiFi network. Since both methods are generally performed in a large time scale, we ignore the diversity of different users, such as different QoS requirements, random channel fading, and user mobility. Instead, we take the average throughput of each user as the performance metric, by averaging among all users and over all channel randomness. This metric has also been used in other related works, such as [6]. In addition, the short-term user scheduling and resource allocation strategies for the LTE network are not considered in this work since we focus on long-term system-level performance. Moreover, since different SBSs have different throughputs, we consider to maximize the minimum average per-user throughput of LTE SBSs to ensure fairness.

On the other hand, for the WiFi network, the minimum per-user throughput should be guaranteed, i.e., R_k^{T} for AP k. This value is very important to ensure the inter-system fairness in such an LTE and WiFi coexisting network, which can be determined by LTE after negotiating with WiFi or monitoring the WiFi network performance.

Let γ_{mi}^{S} be the signal-to-interference-plus-noise ratio (SINR) on the i-th resource block (RB) of SBS m on the licensed spectrum, which can be expressed as

$$\gamma_{mi}^{\mathrm{S}} = \frac{g_{mi}\,P}{\displaystyle\sum_{m'\neq m} g_{m'mi}\,P + \sigma_{\mathrm{S}}^2} \tag{3.4}$$

where g_{mi} is the channel power gain, P is the transmit power, σ_{S}^2 is the noise power, and $g_{m'mi}$ is the interference power gain from SBS m' to SBS m on the i-th RB. Since this chapter emphasizes on the analysis of coexistence scenario between LTE and WiFi, we focus on the average throughput of SBS m on the licensed spectrum, which can be expressed as

$$C_m^{\mathrm{S}} = B_m^{\mathrm{S}} E_m \left\{ \log_2 \left(1 + \gamma_{mi}^{\mathrm{S}} \right) \right\} \tag{3.5}$$

where E_m denotes the expectation over different RBs, channel fading, and interference. It is a bit difficult to average the SINR in Eq. (3.4) due to complicated inter-cell interference. Fortunately, owing to the tractability of the PPP model used in our work, a closed-form expression of the expected overall throughput of a SBS can be derived by utilizing stochastic geometry (please refer to [18, 19] for the detailed derivations, which are omitted here for brevity).

We shall also note that since we focus on the average channel capacity in Eq. (3.5), the channel dynamics would not affect our analysis too much. On the other hand, the short-term channel varying can be dealt with by other radio resource management strategies, such as power control, adaptive modulation, and coding, and have been addressed to a great extent in other literature. Nevertheless, our algorithms do not rely on the detailed interference cancelation and resource management strategies since C_m^{S} is regarded as a constant in the sequel.

Similarly, if the whole unlicensed spectrum is used by SBS m, the expected overall throughput on it can be expressed as

$$C_m^{\mathrm{A}} = B E_m \left\{ \log_2 \left(1 + \gamma_{mi}^{\mathrm{A}} \right) \right\} \tag{3.6}$$

where γ_{mi}^{A} is the SINR of the i-th RB on the unlicensed band, which can be expressed in a similar way as Eq. (3.4). Here we assume that the same LTE transmission protocol in the licensed band is employed in the unlicensed band. We also assume that the average throughputs on different unlicensed bands are the same to each SBS.

3.1.2.4   Problem Formulation

In the following, we will formulate the optimization problem in each method:

(1) Traffic Offloading: Although the choice of which users to offload will affect the system performance, the overall system throughput significantly depends on the total amount of offloaded users in a large time scale. Assuming that SBS m will offload N_{mk} users to AP k, the optimization problem to maximize the minimum average per-user throughput among all SBSs can be formulated as

$$\max_{\{N_{mk}\}} \min_{m} \left\{ \frac{C_m^S}{N_m^S - \sum_{k=1}^{K} N_{mk}} \right\} \tag{3.7}$$

subject to

$$\frac{R\left(N_k^A + \sum_{m=1}^{M} N_{mk}\right)}{N_k^A + \sum_{m=1}^{M} N_{mk}} \geq R_k^T, \ \forall k, \tag{3.7a}$$

$$N_{mk} \leq N_{mk}^{\max}, \ \forall m, k \tag{3.7b}$$

where Eq. (3.7a) ensures the minimum per-user throughput of WiFi and Eq. (3.7b) limits the maximum number of users that can be offloaded from SBS m to AP k. Here, $R(.)$ in Eq. (3.7a) is expressed in Eq. (3.3). Note that we do not count those cellular users offloaded to the WiFi network when calculating the average per-user LTE throughput since they are now connected with WiFi APs and their performance could be guaranteed by the WiFi network.

(2) Resource Sharing: In the resource sharing method, SBS m can occupy L_{mk} time slots from AP k. We aim to obtain the optimal L_{mk} that maximizes the minimum average per-user LTE throughput while guaranteeing the WiFi performance, which can be formulated as

$$\max_{\{L_{mk}\}} \min_{m} \left\{ \frac{C_m^S + C_m^A \sum_{k=1}^{K} L_{mk}}{N_m^S} \right\} \tag{3.8}$$

subject to

$$\frac{R\left(N_k^{\mathrm{A}}\right)\left(1 - \sum_{m=1}^{M} L_{mk}\right)}{N_k^{\mathrm{A}}} \geq R_k^{\mathrm{T}}, \ \forall k \qquad (3.8a)$$

(3) The Hybrid Method: To further improve the LTE throughput, the hybrid method can be utilized, where SBS m offloads N_{mk} users to AP k and occupies its L_{mk} time slots as well. The optimization problem can be modeled as

$$\max_{\{N_{mk}, L_{mk}\}} \min_{m} \left\{ \frac{C_m^{\mathrm{S}} + C_m^{\mathrm{A}} \sum_{k=1}^{K} L_{mk}}{N_m^{\mathrm{S}} - \sum_{k=1}^{K} N_{mk}} \right\} \qquad (3.9)$$

subject to

$$\frac{R\left(N_k^{\mathrm{A}} + \sum_{m=1}^{M} N_{mk}\right)\left(1 - \sum_{m=1}^{M} L_{mk}\right)}{\left(N_k^{\mathrm{A}} + \sum_{m=1}^{M} N_{mk}\right)} \geq R_k^{\mathrm{T}}, \ \forall k, \qquad (3.9a)$$

$$N_{mk} \leq N_{mk}^{\max}, \ \forall m, k \qquad (3.9b)$$

Note that the hybrid method is more general than both traffic offloading and resource sharing. It will become the traffic offloading method if setting $L_{mk} = 0$, $\forall m, k$, and the resource sharing method if setting $L_{mk} = 0$, $\forall m, k$. On the other hand, since L_{mk} and N_{mk} can be jointly optimized, the hybrid method outperforms the other two methods.

3.1.3 Mobile Data Application in Single-BS Scenario

In this section, we will first investigate a simple scenario with only one SBS and one AP to get some insights into the performance comparison of the three methods mentioned above. In the single-BS scenario, we can omit the subscripts "k" and "m" for brevity.

3.1.3.1 Traffic Offloading

In the single-BS scenario, we can rewrite Eq. (3.7a) as

$$\frac{R\left(N^{A} + N\right)}{N^{A} + N} \geq R^{T} \tag{3.10}$$

where N is the number of offloaded users from LTE to WiFi. Since $R(n)$ in (3.3) monotonously decreases with n while the objective function in (3.7) increases with n, we can easily obtain the maximum average per-user throughput of LTE system as

$$\frac{C^{S}}{N^{S} - \min\{N^{*}, N^{\max}\}} \tag{3.11}$$

where N^{*} is the largest integer to satisfy Eq. (3.10). Note that the above result presents the throughput limit of the LTE system with the minimum per-user throughput of the WiFi network guaranteed in the traffic offloading method.

3.1.3.2   Resource Sharing

In the resource sharing method, we assume LTE occupies L ($L \in [0, 1]$) time slots on the unlicensed band. The per-user throughput of WiFi should be guaranteed

$$\frac{R\left(N^{A}\right)(1 - L)}{N^{A}} \geq R^{T} \tag{3.12}$$

Since the left side of Eq. (3.12) decreases but the objective function in Eq. (3.8) increases with L, the optimal L^{*} can be expressed as $L^{*} = 1 - \frac{R^{T}N^{A}}{R(N^{A})}$. Thus, the maximum average per-user LTE throughput can be expressed as

$$\frac{C^{S} + C^{A}L^{*}}{N^{S}} \tag{3.13}$$

This is the throughput limit of the LTE system under the minimum performance requirement of the WiFi network in the resource sharing method.

3.1.3.3   The Hybrid Method

In the hybrid method, the optimization problem can be rewritten as

$$\max_{\{L, N\}} \frac{C^{S} + LC^{A}}{N^{S} - N} \tag{3.14}$$

subject to

$$(1 - L) \cdot \frac{R\left(N^{\mathrm{A}} + N\right)}{N^{\mathrm{A}} + N} \geq R^{\mathrm{T}} \tag{3.14a}$$

Since the objective function increases with both L and N, but the left side of Eq. (3.14a) decreases with them, the equality in Eq. (3.14a) should be achieved. Therefore, we have

$$L = 1 - \frac{R^{\mathrm{T}} \cdot \left(N^{\mathrm{A}} + N\right)}{R\left(N^{\mathrm{A}} + N\right)} \tag{3.15}$$

Then, Eq. (3.14) is equivalent to

$$\max_{0 \leq N \leq N^{\max}} f(N) = \frac{C^{\mathrm{S}} + C^{\mathrm{A}} - \frac{C^{\mathrm{A}} R^{\mathrm{T}} \cdot (N^{\mathrm{A}} + N)}{R(N^{\mathrm{A}} + N)}}{N^{\mathrm{S}} - N} \tag{3.16}$$

which presents the maximum average per-user throughput of the LTE system in the hybrid method.

3.1.3.4 Performance Comparison

Now we address the following two questions as raised in the Introduction part:

1. Traffic offloading and resource sharing: which one is better?
2. How much performance gain can be achieved by the hybrid method?

To answer these questions, we need to assume that $N^{\max}$ is large enough. Then, we introduce the following theorems to address the first problem.

Theorem 1 *When the number of WiFi users, N^{A}, and the number of cellular users, N^{S}, satisfy Eq. (3.17), traffic offloading performs better than resource sharing*

$$N^{\mathrm{A}} < N^{\mathrm{T}} - \frac{C^{\mathrm{A}}}{C^{\mathrm{S}} + C^{\mathrm{A}}} N^{\mathrm{S}} \tag{3.17}$$

where N^{T} is the maximal number of WiFi users satisfying $R\left(N^{\mathrm{T}}\right)/N^{\mathrm{T}} \geq R^{\mathrm{T}}$.

Remark 1 Theorem 1 shows that traffic offloading is better only if the number of existing users in WiFi is small, which can be intuitively explained in the following. When the traffic load of the WiFi network is light, the per-user throughput of WiFi will be greater than that of LTE. In this situation, more LTE users can be offloaded to WiFi networks, leading to better average per-user LTE throughput than in the resource sharing method. However, when N^{A} is large, offloading new users will significantly increase the collision, leading to the degradation of WiFi throughput. Therefore, in this case, it is better to use the unlicensed bands by

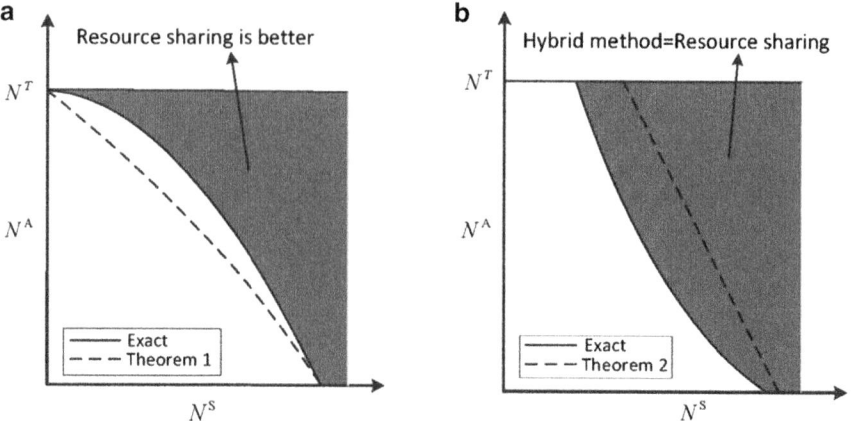

Fig. 3.2 Performance comparison of the three methods in the single-BS scenario. (**a**) Comparison between traffic offloading and resource sharing. (**b**) Comparison between resource sharing and the hybrid method

directly occupying several time slots. Moreover, from the theorem, for larger N^{S} and C^{A}, resource sharing is more likely to be superior. Once N^{S} is large enough, i.e., $N^{\mathrm{S}} > \frac{N^{\mathrm{T}}(C^{\mathrm{S}}+C^{\mathrm{A}})}{C^{\mathrm{A}}}$, resource sharing will be definitely better than traffic offloading regardless of N^{A}. The reason is that more spectrum resource is required to support a large number of LTE users in this case. Figure 3.2 compares the curve in Eq. (3.17) (the dashed line) with the exact boundary curve (the solid line), where resource sharing is superior in the shadowing area. This figure demonstrates that Eq. (3.17) is a necessary condition for traffic offloading surpassing resource sharing. To answer the second question, we present the following theorem.

Theorem 2 *When the number of users in WiFi AP, N^{A}, is sufficiently large, i.e., both the following inequations are satisfied, the hybrid method has the same performance as the resource sharing method.*

$$N^{\mathrm{A}}N^{\mathrm{S}}\beta + \frac{N^{\mathrm{S}} + N^{\mathrm{A}}}{R\left(N^{\mathrm{A}}\right)} \geq \frac{C^{\mathrm{S}} + C^{\mathrm{A}}}{C^{\mathrm{A}}R^{\mathrm{T}}}, \tag{3.18}$$

$$\left(N^{\mathrm{S}} + N^{\mathrm{A}} - N^{\mathrm{T}}\right)N^{\mathrm{T}}\beta + \frac{N^{\mathrm{S}} + N^{\mathrm{A}}}{R\left(N^{\mathrm{T}}\right)} \geq \frac{C^{\mathrm{S}} + C^{\mathrm{A}}}{C^{\mathrm{A}}R^{\mathrm{T}}} \tag{3.19}$$

where $\beta = \min\limits_{N}\left\{\frac{1}{R(N^{\mathrm{A}}+N+1)} - \frac{1}{R(N^{\mathrm{A}}+N)}\right\}$

Remark 2 Theorem 2 indicates that when N^{A} is large enough, offloading users to WiFi is no longer necessary and the hybrid method is identical to the resource sharing method. The reason can be explained in a similar way as in Theorem 1.

Note that to guarantee the minimum per-user throughput of WiFi system, $N^A < N^T$ should be satisfied. However, for a sufficiently large N^S, there will always exist N^A satisfying (3.18) and (3.19) since the left sides of both inequations increase with N^S and N^A. Moreover, as indicated by the above inequations, the dynamic region of N^A increases with N^S, which shows that traffic offloading is more likely to be unnecessary for large N^S and the hybrid method can be degraded into the resource sharing method in this case. Figure 3.2 demonstrates the theorem, where the solid line corresponds to the exact boundary line, the dashed line is given by Eqs. (3.18) and (3.19), and the shadowing area means that the hybrid method has the same performance as the resource sharing method. From the figure, Eqs. (3.18) and (3.19) serve as necessary conditions.

3.1.3.5 Extension to Downlink

In the downlink case, the saturation throughput in (3.3) would be a constant since there is no competing users, that is,

$$R(n) = R, \forall n \tag{3.20}$$

where R is the overall WiFi throughput. In this case, the maximum supported WiFi user $N^T = \frac{R}{R^T}$. Therefore, if Eq. (3.17) exists, we can also demonstrate that

$$\frac{C^S}{N^S + N^A - N^T} > \frac{C^S + C^A}{N^S} > \frac{C^S + C^A - \frac{N^A}{N^T}C^A}{N^S} \tag{3.21}$$

which means that traffic offloading is better than resource sharing.

On the other hand, $\beta = 0$ in this case. Therefore, Eqs. (3.18) and (3.19) reduce to

$$\frac{N^S + N^A}{R} \geq \frac{C^S + C^A}{C^A R^T} \tag{3.22}$$

From Eq. (3.22), we can easily derive that

$$\max_{N} f(N) = f(0) = \frac{C^S + C^A - \frac{N^A}{N^T}C^A}{N^S} \tag{3.23}$$

which implies that the hybrid method has the same performance as resource sharing in this case (Fig. 3.3).

From the above discussion, our derivation can also be extended into the downlink case.

Fig. 3.3 Performance comparison in the single-BS scenario. (**a**) LTE throughput with different numbers of cellular users. $N^A = 4$. (**b**) LTE throughput with different numbers of WiFi users. $N^S = 35$

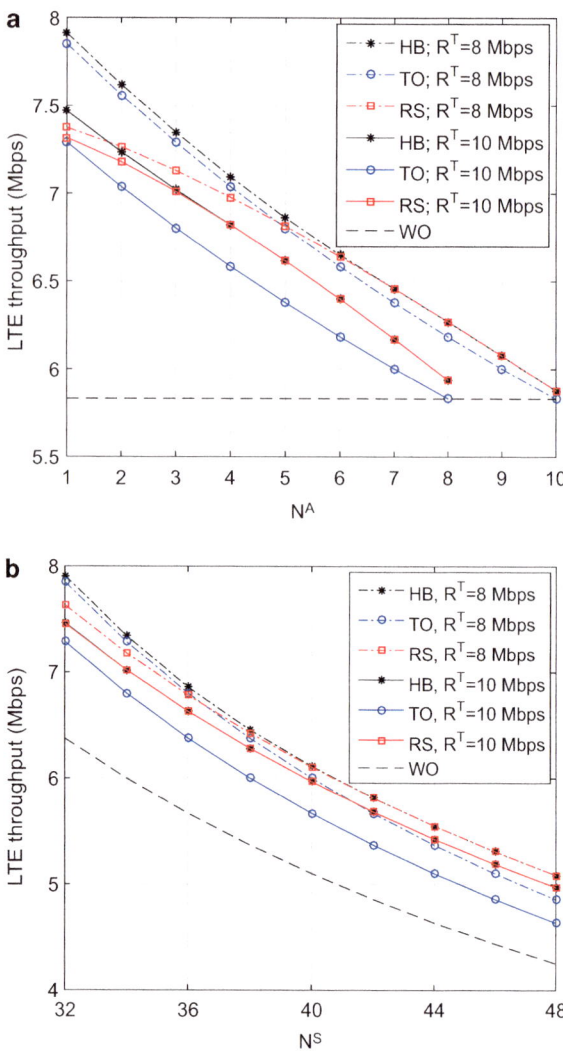

3.1.4   Mobile Data Application in Multiple-BS Scenario

In this section, we will discuss the three methods in the multi-BS scenario, i.e., with multiple SBSs and multiple APs. We first derive the performance upper bound and then develop practical algorithm for each method.

3.1.4.1 Traffic Offloading

First, we will investigate the optimal traffic offloading problem in (3.7). We denote RTO as the minimum average per-user throughput among SBSs, i.e., $R^{TO} = \min_{m} \left\{ \frac{C_m^S}{N_m^S - \sum_{k=1}^{K} N^{mk}} \right\}$. Then, by parametric algorithm [20], (3.7) can be transformed into

$$\max R^{TO} \tag{3.24}$$

subject to

$$\frac{C_m^S}{N_m^S - \sum_{k=1}^{K} N_{mk}} \geq R^{TO}, \forall m, \tag{3.24a}$$

$$\frac{R\left(N_k^A + \sum_{m=1}^{M} N_{mk}\right)}{N_k^A + \sum_{m=1}^{M} N_{mk}} \geq R_k^T, \forall k, \tag{3.24b}$$

$$N_{mk} \leq N_{mk}^{\max}, \forall m, k \tag{3.24c}$$

The above problem is an integer optimization problem and is not easy to solve. To make it better tractable, we first relax N_{mk} into a continuous variable and further ignore the constraint in Eq. (3.24c) to achieve its upper bound. With these assumptions, we can obtain a closed-form solution.

Moreover, it is rather intuitive that to achieve the largest R^{TO}, more users should be offloaded to WiFi since $R(n)$ is a decreasing function and R^{TO} increases with N_{mk}. In this situation, Eq. (3.24b) can be rewritten as $\sum_{m=1}^{M} N_{mk} = N_k^T - N_k^A$, where N_k^T denotes the maximum number of users that could be supported in AP k to satisfy the minimum per-user throughput, as $R\left(N_k^T\right)/N_k^T = R_k^T, \forall k$.

With the continuous relaxation of N_{mk} and the above analysis, an upper bound for R^{TO} could be expressed in a closed-form expression as shown in the following corollary.

Corollary 1 *An upper bound for R^{TO} can be expressed as*

$$R^{TO} = \frac{\sum_{m=1}^{M} C_m^S}{\sum_{m=1}^{M} N_m^S + \sum_{k=1}^{K} N_k^A - \sum_{k=1}^{K} N_k^T} \tag{3.25}$$

Algorithm 1 The algorithm for traffic offloading

1: Initialize $N_{mk} = 0, \forall m, k, S = \{1, 2, \ldots, M\}, \Omega_m = \{k | N_{mk} < N_{mk}^{\max}$ and $\sum_{m=1}^{M} N_{mk} < N_k^{T} - N_k^{A}\}$

2: Do

3:    Select $m^* = \underset{m \in S}{\operatorname{argmin}} \dfrac{C_m^S}{N_m^S - \sum_{k=1}^{K} N_{mk}}$.

4:    If $\Omega_{m^*} = \phi$

5:       $S = S - \{m^*\}$.

6:    Else

7:       Select $k^* = \underset{k \in \Omega_{m^*}}{\arg \max} \left(N_k^{T} - N_k^{A} - \sum_{m=1}^{M} N_{mk} \right)$.

8:       $N_{m^*k^*} = N_{m^*k^*} + 1$.

9:       Update Ω_{m^*}.

10:   End if

11: Until $S = \phi$

Algorithm 2 The algorithm for resource sharing

1: Initialize $L_{mk} = 0, S = \{1, 2, \ldots, M\}, R_k = R\left(N_k^A\right)/N_k^A, \Omega_m = \{k | (1 - \sum_{m=1}^{M} L_{mk}) R_k \geq R_k^{T}$ and $a_{mk} = 1\}$.

2: Do

3:    Select $m^* = \underset{m \in S}{\operatorname{argmin}} \dfrac{C_m^S + C_m^A \sum_{k=1}^{K} L^{mk}}{N_m^S}$.

4:    If $\Omega_{m^*} = \phi$

5:       $S = S - \{M^*\}$.

6:    Else

7:       Select $k^* = \underset{k \in \Omega_{m^*}}{\arg \max} \left\{ \left(1 - \sum_{m=1}^{M} L_{mk}\right) R_k - R_k^{T} \right\}$.

8:       $L_{m^*k^*} = L_{m^*k^*} + 1/L^A$.

9:       Update Ω_{m^*}.

10:   End if

11: Until $S = \phi$

It should be noted that the above equation is achieved with the assumption of continuous N_{mk} and the relaxation of constraint equation (3.24c). In the following, we present a practical traffic offloading algorithm in the multi-BS scenario, as described in Algorithm 1. The main idea of the algorithm is to offload one user from the SBS with the minimum average per-user throughput to an overlapped AP which has the largest capacity in each iteration until all APs reach the minimum per-user throughput threshold.

3.1.4.2   Resource Sharing

Similar to the traffic offloading case, the problem in Eq. (3.8) can be reformulated as

$$\max R^{\mathrm{RS}} \tag{3.26}$$

Algorithm 3 The algorithm for the hybrid method

1: Initialize $N_{mk} = 0, L_{mk} = 0, \forall m, k, S = \{1, 2, \ldots, M\}, R_k = R\left(N_k^A + \sum_{m=1}^{M} N_{mk}\right) / \left(N_k^A + \sum_{m=1}^{M} N_{mk}\right), \Omega_m = \{k | (1 - \sum_{m=1}^{M} L_{mk}) R_k \geq R_k^T$ and $a_{mk} = 1\}$.
2: Do
3:   Select $m^* = \underset{m \in S}{\arg\min} \dfrac{C_m^S + C_m^A \sum_{k=1}^{K} L_{mk}}{N_m^S - \sum_{k=1}^{K} N_{mk}}$.
4:     If $\Omega_{m^*} = \phi$
5:       $S = S - \{m^*\}$.
6:     Else
7:       Select $k^* = \underset{k \in \Omega_{m^*}}{\arg\max} \left\{\left(1 - \sum_{m=1}^{M} L_{mk}\right) R_k - R_k^T\right\}$.
8:       Calculate R^{TO} and R^{RS} according to (3.25), (3.30).
9:       If $R^{TO} > R^{RS}$ and $N_{m^*k^*} < N_{m^*k^*}^{\max}$.
10:        $N_{m^*k^*} = N_{m^*k^*} + 1$.
11:        Update R_{k^*}
12:      Else
13:        $L_{m^*k^*} = L_{m^*k^*} + 1/L^A$.
14:      End if
15:      Update Ω_{m^*}.
16:    End if
17: Until $S = \phi$

---

subject to

$$\frac{C_m^S + C_m^A \sum_{k=1}^{K} L_{mk}}{N_m^S} \geq R^{RS}, \ \forall m, \tag{3.26a}$$

$$\frac{R\left(N_k^A\right)\left(1 - \sum_{m=1}^{M} L_{mk}\right)}{N_k^A} \geq R_k^T, \ \forall k \tag{3.26b}$$

Again, we can relax L_{mk} into a continuous variable to obtain an upper bound of R^{RS}, as shown in the following corollary.

Corollary 2 *An upper bound for R^{RS} can be expressed as*

$$R^{RS} = \frac{\sum_{m=1}^{M} \dfrac{C_m^S}{C_m^A} + \sum_{k=1}^{K} \left(1 - \dfrac{N_k^A R_k^T}{R(N_k^A)}\right)}{\sum_{m=1}^{M} \dfrac{N_m^S}{C_m^A}} \tag{3.27}$$

We now develop a practical algorithm for the resource sharing method taking into account the discrete time slots, as presented in Algorithm 2. The main idea of the algorithm is similar to the traffic offloading case. In each iteration, we pick up the SBS with the minimum average per-user throughput and allocate one time slot to it from the AP with the largest capacity.

3.1.4.3   The Hybrid Method

Now, we begin to investigate the hybrid method. The optimization problem in Eq. (3.9) can be rewritten as

$$\max R^{\mathrm{HB}} \tag{3.28}$$

subject to

$$\frac{C_m^{\mathrm{S}} + C_m^{\mathrm{A}} \sum\limits_{k=1}^{K} L_{mk}}{N_m^{\mathrm{S}} - \sum\limits_{k=1}^{K} N_{mk}} \geq R^{\mathrm{HB}}, \ \forall m, \tag{3.28a}$$

$$\frac{R \left(N_k^{\mathrm{A}} + \sum\limits_{m=1}^{M} N_{mk} \right) \left(1 - \sum\limits_{m=1}^{M} L_{mk} \right)}{\left(N_k^{\mathrm{A}} + \sum\limits_{m=1}^{M} N_{mk} \right)} \geq R_k^{\mathrm{T}}, \ \forall k, \tag{3.28b}$$

$$N_{mk} \leq N_{mk}^{\max}, \ \forall m, k \tag{3.28c}$$

If we relax both N_{mk} and L_{mk} into continuous variables and ignore the constraint equation (3.28c), we have the following corollary to express an upper bound of the maximum average per-user throughput.

Corollary 3 *An upper bound for R^{HB} can be expressed as*

$$R^{\mathrm{HB}} = \max_{\{N_{mk}\}} \frac{\sum\limits_{m=1}^{M} \frac{C_m^{\mathrm{S}}}{C_m^{\mathrm{A}}} + \sum\limits_{k=1}^{K} \left\{ 1 - \frac{R_k^{\mathrm{T}} \cdot \left(N_k^{\mathrm{A}} + \sum\limits_{m=1}^{M} N_{mk} \right)}{R \left(N_k^{\mathrm{A}} + \sum\limits_{m=1}^{M} N_{mk} \right)} \right\}}{\sum\limits_{m=1}^{M} \frac{N_m^{\mathrm{S}}}{C_m^{\mathrm{A}}} - \sum\limits_{m=1}^{M} \frac{\sum\limits_{k=1}^{K} N_{mk}}{C_m^{\mathrm{A}}}} \tag{3.29}$$

We can develop a joint resource sharing and traffic offloading algorithm for the hybrid method, as shown in Algorithm 3. In each iteration, we pick up the SBS with the minimum average per-user throughput and an overlapped AP with the largest capacity. Then we decide to offload one user to this AP or occupy its one time slot depending on which one is better. Specifically, traffic offloading is selected if $R^{\mathrm{TO}} > R^{\mathrm{RS}}$ and resource sharing is selected otherwise, where R^{TO} is given in

Eq. (3.25) and

$$R^{\text{RS}} = \frac{\sum_{m=1}^{M} \frac{C_m^{\text{S}}}{C_m^{\text{A}}} + \sum_{k=1}^{K} \left(1 - \frac{N_k^{\text{A}} R_k^{\text{T}}}{R\left(N_k^{\text{A}}\right)}\right)}{\sum_{m=1}^{M} \frac{N_m^{\text{S}} - \sum_{k=1}^{K} N_{mk}}{C_m^{\text{A}}}} \qquad (3.30)$$

We can see that all the three algorithms have the computational complexity of only $\mathcal{O}\,(M \cdot N)$. Therefore, they can be easily implemented in practical systems.

3.1.4.4 Performance Evaluation

In this section, numerical results are presented to evaluate the performance of the proposed algorithms. The transmit power at the SBS is 20 dBm and the noise power on both licensed and unlicensed bands is −95 dBm. The path loss model is $15.3 + \alpha \times 10\log_{10}(d)$ dB, where d is in meter and α is 3.75 and 5 for licensed and unlicensed bands, respectively [21]. The LTE SBSs are distributed according to homogeneous PPP and the interference and rate model in [18] is adopted. The WiFi APs are deterministically deployed for given LTE SBS locations and both LTE and WiFi users are uniformly distributed.

The WiFi network is equipped with the IEEE 802.11n protocol working at the 5 GHz band and we adopt the RTS/CTS mechanism in the simulations. Other WiFi parameters are listed in Table 3.1. Here, we assume that the channel data rate of the IEEE 802.11n system is 130 Mbps, which is a typical value for 20 MHz bandwidth. However, similar results can be expected for other channel data rates. In the following figures, TO, RS, and HB stand for traffic offloading, resource sharing, and the hybrid method, respectively, while WO stands for the original SBS system without any of the three methods.

We now present the results for the multi-BS scenario, where four nearby SBSs share unlicensed spectrum with seven WiFi APs, as shown in Fig. 3.4. Specifically, SBS1 is overlapped with AP 1 to AP 3 with maximum offloading numbers of 7, 11, and 10, respectively, SBS2 is overlapped with AP 4 and AP 5, with maximum offloading numbers of 8 and 12, respectively, SBS3 is overlapped with AP 1, 2, and 7, with the same maximum offloading number of 7 to each AP, and SBS4 is overlapped with AP 1, AP 6, and AP 7 with maximum offloading numbers of 8, 8, and 9, respectively. It is assumed that the per-user throughput threshold of each WiFi AP is the same, i.e., $R_k^{\text{T}} = 9$ Mbps, $\forall k$, and each SBS has the same number of cellular users, i.e., $N_m^{\text{S}} = N^{\text{S}}$, $\forall m$. The C_m^{S} of the four SBSs are 213.5, 223.8, 235.2, and 248 Mbps, and the C_m^{A} of the four SBSs are 73.4, 86.3, 100.9, and 117.5 Mbps.

Figure 3.5a illustrates the average per-user throughput of each SBS in the three methods where the dashed lines correspond to the upper bound of each method. From the figure, SBSs can achieve almost the same performance in each method, resulting from the maximization of the minimum average per-user throughput in the objective function. The performance upper bound of each method

Table 3.1 Size of dataset

Parameters	Settings
Noise power	-95 dBm
Path loss model (licensed,unlicensed)	$15.3 + \alpha \times 10\log_{10}(d)\ (\alpha = 3.75, 5)$
Transmit power	20 dBm
$E[P]$	1500 byte
B^S, B	20 MHz, 20 MHz
$CW_{\min}$	16
$CW_{\max}$	1024
R_{limit}	6
WiFi channel bit rate	130 Mbps
PHY header	192 bits
MAC header	224 bits
T_δ	20 μs
SIFS	16 μs
DIFS	50 μs
Slot time	9 μs
ACK	112 bits+PHY header
RTS	160 bits+PHY header
CTS	112 bits+PHY header
L^A	100

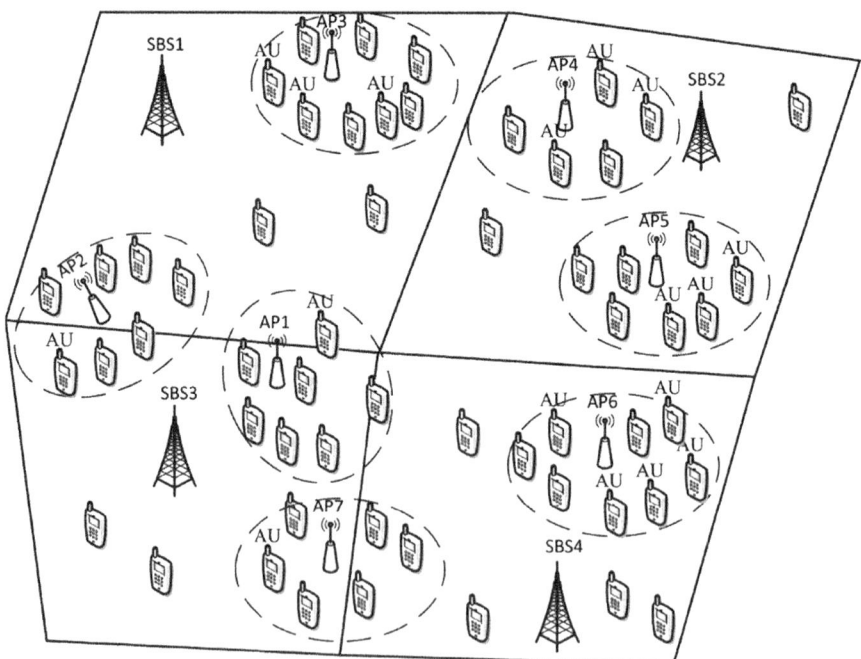

Fig. 3.4 Simulation model for the multi-BS scenario

Fig. 3.5 Average per-user
throughput of each SBS and
WiFi AP. $N^S = 35$. $N^A =$
$\{4, 4, 6, 6, 6, 8, 4\}$. (**a**)
Average per-user throughput
of each SBS. (**b**) Average
per-user throughput of each
SBS

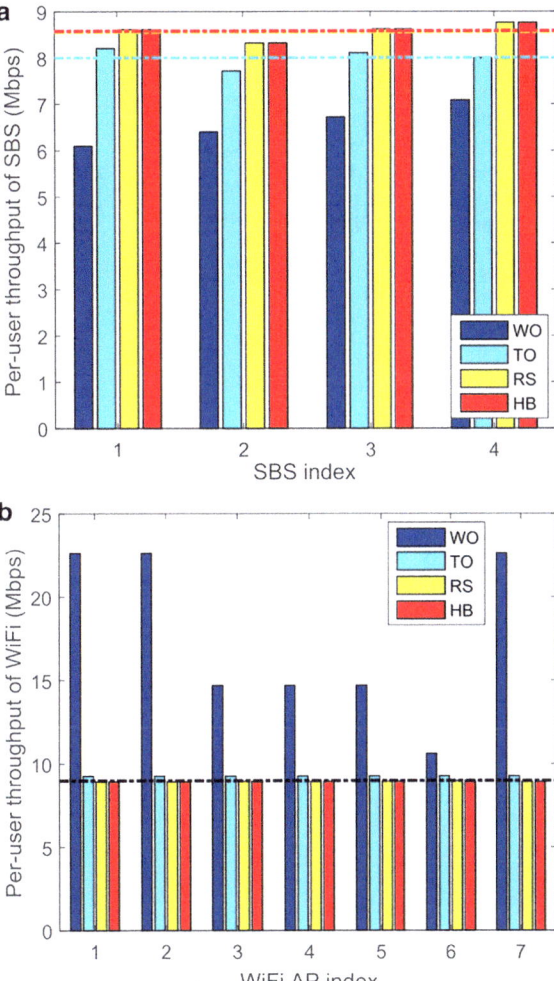

is also demonstrated. The performance of WiFi APs in each method is depicted in
Fig. 3.5b, where the dashed line shows the required minimum per-user throughput.
As predicted, each WiFi AP achieves a per-user throughput that is a little more than
the given threshold.

Figure 3.6a, b presents the minimum average per-user LTE throughput with
different numbers of cellular users and WiFi users, respectively. From Fig. 3.6a,
the hybrid method achieves a better performance than the others when N^S is small,
while it has the same performance as resource sharing for large N^S. Also, traffic
offloading is first superior and then inferior to resource sharing as N^S increases.
These phenomena are the same as in the single-BS scenario as shown in Fig. 3.3a.
A similar result can be found in Fig. 3.6b, where $N_k^A = N^A$, $\forall k$. In both figures, the
upper bounds for different methods in Corollaries 1–3 are validated.

Fig. 3.6 Performance comparison of different methods. (**a**) Minimum average per-user LTE throughput with different numbers of cellular users. $N^A = \{4, 4, 6, 6, 6, 8, 4\}$. (**b**) Minimum average per-user LTE throughput with different numbers of WiFi users. $N^S = 35$

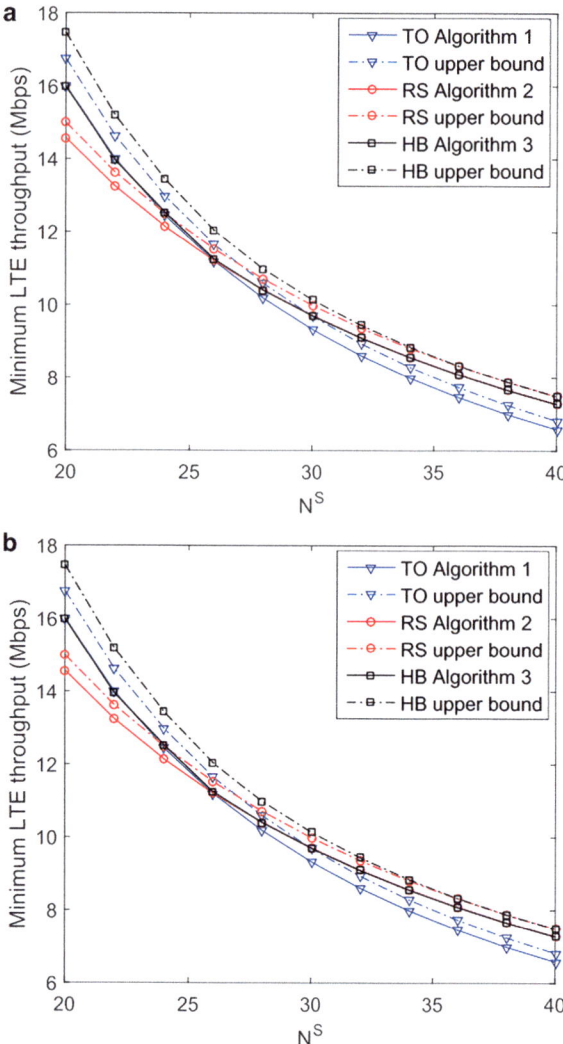

3.2   Mobile Data Offloading and Resource Optimization in LTE and WiFi Systems

Traditional mobile data offloading transfers cellular users to WiFi networks to relieve the cellular system from the pressure of the ever-increasing data traffic load. However, the spectrum utilization of the WiFi network is bound to suffer from potential packet collisions due to its contention-based access protocol, especially when the number of competing WiFi users grows large. To tackle this problem, we propose transferring some WiFi users to be served by the LTE system, in contrast to the traditional mobile data offloading which effectively offloads LTE traffic to the WiFi network. Meanwhile, leveraging the emerging LTE in unlicensed spectrum (LTE-U) technology, some unlicensed spectrum resources may be allocated to the LTE system in compensation for handling more WiFi users. In this way, a win–win situation would be generated since LTE can generally achieve better performance than WiFi due to its capability of centralized coordination. To facilitate it, three important challenging issues are addressed in the chapter: which WiFi users should be transferred; how many WiFi users need to be transferred; and how much unlicensed resources should be relinquished to the LTE-U network. We investigate three different user transfer schemes according to the availability of channel state information (CSI): the random transfer, the distance-based transfer, and the CSI-based transfer. In each scheme, the minimum required amount of unlicensed resources under a given transferred user number is analyzed. Furthermore, we utilize the Nash bargaining solution (NBS) to develop joint user transfer and unlicensed resource allocation strategy to fulfill the win–win situation for both networks, whose performance is demonstrated by numerical simulation.

3.2.1   Transfer WiFi Users to the LTE-U System: Concepts and Applications

The cellular data traffic has dramatically increased in the past few years and it will continue to undergo an 11-fold increase through 2013–2018 [21]. To meet this challenge, mobile network operators have deployed many small cell base stations (SBSs) and WiFi access points (APs) to offload the cellular traffic, which is referred to as mobile data offloading. Due to its low cost and license-exemption, the deployment of WiFi APs can remarkably increase the network capacity without incurring too much operational and capital expenditures. Therefore, the vast majority of recent research is focusing on offloading cellular traffic to the WiFi network [1–10].

However, since most of the WiFi APs use the distributed coordination function (DCF) as the media access control (MAC) layer protocol, the spectrum utilization of the WiFi network is bound to suffer from potential packet collisions, especially when there are too many competing WiFi users. To find a more effective unlicensed spectrum utilization method and alleviate the spectrum scarcity problem of cellular networks, major cellular operators and vendors have launched the investigation on LTE in unlicensed spectrum, known as LTE-U or licensed-assisted access (LAA) in the 3GPP framework [22–24]. In general, LTE-U can achieve better unlicensed spectrum utilization than WiFi due to its centralized scheduling and other advanced physical and MAC layer techniques [25].

The major challenge of implementing LTE-U is how to fairly and harmoniously coexist with the WiFi network deployed in the same unlicensed spectrum. It has been demonstrated that LTE-U would be a better neighbor to WiFi than an additional WiFi network if its transmission is carefully controlled [11]. To this end, the listen-before-talk (LBT) protocol [26] and the duty-cycle method [27, 28] have been developed to achieve a fair coexistence between the two networks. More recently, non-orthogonal resource sharing by leveraging multiple antennas in LTE and WiFi and utilizing different frame structures in different systems has been developed in [28]. Other coexisting mechanisms rely on an inter-system coordinator that manages the transmission of both LTE-U and WiFi networks. A logically centralized optimization framework has been developed for WiFi and LTE coexistence, which involves dynamic spectrum management and inter-network coordination [29]. In [30], optimal resource allocation algorithms for both dual-band femtocell systems and integrated femto-WiFi networks have been developed. Joint licensed and unlicensed resource allocation algorithms for LAA systems have been developed in [13] and [31] for throughput and energy efficiency maximization, respectively. In [32], we have compared the performance of traffic offloading and LTE-U in terms of large-scale system throughput and developed a novel mechanism to exploit the advantages of both traffic offloading and LTE-U.

Nowadays, a large number of WiFi devices (e.g., smartphones) are also equipped with cellular capabilities and this number will keep increasing in the future. Furthermore, LTE-U can provide a new and effective way to utilize the unlicensed spectrum that has been used by WiFi systems. As a result, we should rethink the traditional mobile data offloading, which generally offloads the LTE users to the WiFi network. In this chapter, we propose to transfer WiFi users to the LTE-U system, which is just opposite to the existing mobile offloading schemes [1–10], and at the same time relinquish some unlicensed resources to the LTE-U system to support the transferred users. Since LTE-U can achieve a higher spectral efficiency than WiFi on the unlicensed spectrum, the quality of service (QoS) of those transferred users can be better ensured. Meanwhile, the performance of the WiFi network could be improved owing to the decreased number of competing users. Furthermore, due to more efficient resource management in LTE-U, less unlicensed spectrum will be required by LTE-U to serve those transferred users than by WiFi and the leftover unlicensed spectrum can be used to improve the QoS of the existing cellular users. Therefore, a win–win situation for the transferred users, the remaining WiFi users, and the LTE users can be achieved by this strategy.

However, there exist some challenging issues to fulfill such a win–win strategy. First, the number of transferred WiFi users and the amount of unlicensed resources relinquished to the LTE-U system should be carefully determined. In general, more unlicensed resources should be allocated to the LTE-U system if more WiFi users are transferred. However, this might degrade the performance of the WiFi network since few unlicensed resources would remain. On the other hand, transferring fewer WiFi users will not substantially improve the WiFi performance since the number of competing WiFi users could still remain high. Besides, which users should be transferred is also an important issue.

To address the above challenges, we first propose three different user transfer schemes according to the availability of channel state information (CSI). They are the random transfer (RT), the distance-based transfer (DT), and the CSI-based transfer (CT). The minimum required amount of unlicensed time slots under a given transferred user number will be first analyzed for each scheme. The number of users who need to be transferred and the amount of unlicensed spectrum that needs to be relinquished to the LTE-U system should be jointly determined by negotiation between the two networks. The challenge here is that WiFi and LTE-U have contradicting intentions, i.e., WiFi intends to transfer more users to LTE-U and relinquish as less unlicensed spectrum as possible, while the expectation of LTE-U is just opposite. Therefore, to achieve an effective balance/tradeoff, we utilize the Nash bargaining solution (NBS) to develop a joint user transfer and unlicensed resource allocation algorithm.

The main contributions of this chapter are summarized as follows:

- Opposite to the traditional mobile data offloading, we propose to transfer WiFi users to the LTE-U system that is allocated with certain amount of unlicensed spectrum at the same time for compensation. The proposed scheme can create a win–win situation for both networks, i.e., the average per-user throughput of the WiFi system can be improved and the LTE-U system can gain extra unlicensed spectrum to better serve the cellular users as well.
- Three different user transfer schemes, namely the random transfer, the distance-based transfer, and the CSI-based transfer, are investigated. The minimum required amount of unlicensed time slots under a given transferred user number is derived for each scheme.
- To achieve the win–win strategy, we develop a joint user transfer and unlicensed resource allocation algorithm based on the NBS. The closed-form expressions for the unlicensed time slot allocation are derived in both single-AP and multi-AP scenarios. We also develop an effective algorithm to determine the optimal number of transferred users in each AP.

The rest of this chapter is organized as follows: In Sect. 3.1.2, the system model is described and three different user transfer schemes are introduced. The minimum required amount of unlicensed time slots for a given transferred user number is analyzed in Sect. 3.1.3. The joint user transfer and unlicensed resource allocation algorithms to achieve the win–win strategy in the single-AP and the multi-AP scenarios are developed in Sect. 3.1.4.

3.2.2 A Deterministic Location Model for the WiFi APs

As illustrated in Fig. 3.7, we consider a scenario with one LTE SBS and K WiFi APs operating at the same unlicensed spectrum. In this chapter, a deterministic location model for the WiFi APs is adopted, i.e., the distance between the k-th WiFi AP and the SBS is denoted as L_k, which is a deterministic value and is known for resource allocation. We further assume that the coverage areas of the WiFi APs are not overlapped and each transmits on an orthogonal channel in the 5 GHz unlicensed spectrum. Thus, there is no interference among different WiFi APs. This model has also been adopted in other literatures, such as [5, 30].

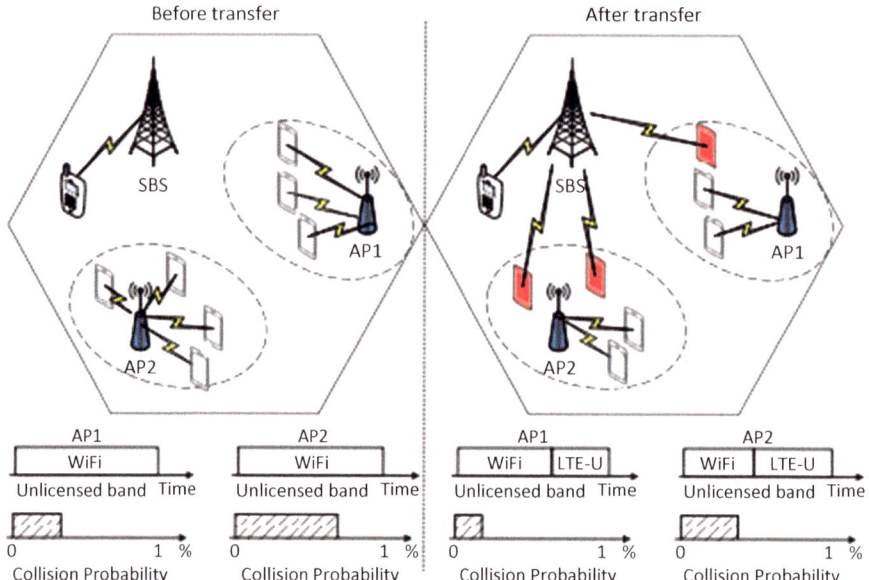

Fig. 3.7 System model for LTE and WiFi coexistence

Denote I as the number of cellular users in the SBS, which are working at the licensed spectrum. The number of WiFi users in the k-th AP is N_k. The IEEE 802.11n protocol is used for WiFi transmission. We assume that WiFi users are also equipped with LTE transceivers. Those devices with only WiFi transceivers in practice can be regarded as non-offloadable devices and therefore could be ignored without affecting our analysis.

To facilitate the coexistence of WiFi and LTE, we assume that there exists an inter-system coordinator, which performs the WiFi user transfer and unlicensed resource allocation, as in [13, 33]. We shall note that our proposal is very useful for the case that LTE-U and WiFi are deployed by the same network operator. In this case, the inter-system coordinator can be implemented by the cellular network operator itself. Otherwise, it can be implemented by a third-party vendor that provides service enhancement for both WiFi and LTE. Furthermore, the time period in the unlicensed spectrum will be divided into several time slots: some for WiFi and the others for LTE-U. This can be accomplished by the duty-cycle method [27, 28] as controlled by the inter-system coordinator.

3.2.2.1 WiFi Model

We assume that the WiFi system is working in the saturation state, i.e., each WiFi station always has packets to transmit. It is noted that our work can also be extended into non-saturation systems, its saturation throughput is a function of the number of competing stations, l. Let P_{tr} and P_s be the probabilities that there is at least one transmit signal in a time slot and that there is no collision in a channel, respectively. Then, from [15]

$$P_{tr} = 1 - (1 - \tau)^l , \tag{3.31}$$

$$P_s = l\tau (1 - \tau)^{l-1} / P_{tr} \tag{3.32}$$

where τ is the transmission probability of each user. According to [15], the per-user saturation throughput of the WiFi network can be expressed as

$$R(l) = \frac{P_{tr} P_s E[P] l^{-1}}{(1 - P_{tr}) T_\sigma + P_{tr} P_s T_s + P_{tr} (1 - P_s) T_c} \tag{3.33}$$

where T_s is the average time the channel is sensed busy because of a successful transmission, T_c is the average time the channel is sensed busy by each station during a collision, T_σ is the duration of an empty time slot, and $\mathbb{E}[P]$ is the average packet size.

In the above, T_s and T_c depend on the specific mechanism, i.e., the basic mechanism or the RTS/CTS mechanism, as well as the channel data rate [15, 32]. Both T_s and T_c decrease with the data rate, leading to the increase of network throughput. Future WiFi, like IEEE 802.11ac, may have comparable or even larger data rate than LTE. However, the network throughput could not be progressively increased mainly due to the contention-based MAC protocol. As we can see from Eq. (3.33), the overall network throughput decreases with the user number, l. Thus, to increase the network throughput, it is reasonable to transfer WiFi users to the LTE-U network.

In this chapter, we focus on uplink traffic transfer since uplink is the main bottleneck of WiFi [34]. The WiFi uplink uses contention-based DCF protocol, whose performance will deteriorate severely as the number of uplink users increases. Therefore, transferring uplink traffic to the LTE-U network can reduce the number of competing users and consequently improve the WiFi throughput more significantly than transferring downlink traffic.

We assume that n_k WiFi users need to be transferred to the LTE-U network from the k-th AP. These transferred users will transmit in the LTE-U network. In order to keep the QoS of the existing cellular users unaffected, we assume that an amount of ρ_k unlicensed time slots will be relinquished from the k-th WiFi AP to the LTE SBS to support those transferred users. Here, we assume the total amount of unlicensed time slots is normalized and hence $0 \le \rho_k \le 1$.

After user transfer and unlicensed time slot relinquishing, the number of remaining users and the remaining amount of time slots in the k-th WiFi AP are $N_k - n_k$ and $1 - \rho_k$, respectively. Therefore, the per-user saturation throughput of the k-th WiFi AP becomes $(1 - \rho_k) R(N_k - n_k)$.

3.2.2.2 LTE-U Model and User Transfer Scheme

As mentioned above, the unlicensed spectrum will be used to serve those WiFi users transferred to the LTE-U network. If WiFi user i communicates with the LTE SBS, its signal-to-noise ratio (SNR) can be expressed as

$$\gamma_i = \frac{P_i H_i d_i^{-\alpha_u}}{\omega} \tag{3.34}$$

where P_i is the transmission power of user i, d_i is the distance between the SBS and WiFi user i, α_u is the path loss exponent on the unlicensed spectrum, H_i is a random variable capturing the fast fading between the SBS, and $\omega = N_0/C_u$ with C_u being the path loss constant and N_0 being the background noise on a resource block. We assume H_i varies according to independent and identically distributed (i.i.d) Rayleigh fading, i.e., $H_i \sim \text{Exp}(3.31)$, and thus $Pr\,(H_i \geq x) = e^{-x}, \forall i$. Here, we ignore the effect of log-normal shadowing for simplicity and tractability. As indicated in [35, 36], the randomness of user locations can approximately simulate shadowing: as the variance of log-normal shadowing increases, the resulting propagation losses between the users and the SBS will converge to those in a Poisson distributed network.

Moreover, we only consider the single-cell case in the above equation. However, our analysis can be easily extended into the multi-cell case if regarding multi-cell interference as noise. This is reasonable since the power of inter-cell interference in the 5 GHz unlicensed bands is much weaker than the signal power due to large channel attenuation on high frequency spectrum.

The following three user transfer schemes will be investigated:

- Random transfer (RT): In this scheme, the inter-system coordinator exploits no information on the WiFi users. Therefore, it randomly selects a certain number of WiFi users to transfer.
- Distance-based transfer (DT): In this scheme, the inter-system coordinator knows the distance between each WiFi user and the LTE SBS, i.e., d_i. This can be realized by the global positioning system (GPS) module embedded in the WiFi device, e.g., the WiFi user sends its GPS coordination to the inter-system coordinator. If there is no active GPS module, the SBS can estimate the distance according to the average power of the WiFi signal and then feed it back to the inter-system coordinator.
- CSI-based transfer (CT): In this scheme, both d_i and H_i are known to the inter-system coordinator, which can be realized by the SBS that estimates the instantaneous CSI of the WiFi signal and then feeds it back to the inter-system coordinator.

Among the three schemes mentioned above, the RT is the simplest. However, its performance is the worst since it exploits no information for user transfer. The CT can achieve the best performance; however, it bears the highest complexity since the exact CSI is not easy to obtain realistically. In fact, DT is the most useful scheme for the practical system as it could achieve a good performance and only requires the distance information of each WiFi user, which is easy to be acquired in current wireless systems. Therefore, the CT scheme can be regarded as a performance upper bound, while the RT scheme can be regarded as a benchmark. Through numerical simulation, we can observe that DT has a much better performance than RT, while it performs almost the same as CT.

3.2.2.3 Problem Description

After acquiring the information, the inter-system coordinator needs to determine how many users in each WiFi AP should be transferred, i.e., n_k, which users need to be transferred, and how much time slots should be relinquished from the k-th WiFi AP to the LTE SBS, i.e., ρ_k. As mentioned before, this procedure can bring mutual benefits for both LTE and WiFi users since LTE can better utilize the unlicensed spectrum. To be more specific, the average per-user throughput in the WiFi AP after user transfer can be improved due to the reduced number of competing WiFi users. On the other hand, the amount of the relinquished unlicensed time slots could be larger than the minimum required amount of time slots to support the transferred WiFi users in the LTE-U network. Therefore, the leftover unlicensed time slots can be used to improve the QoS of the existing LTE users.

Based on the above discussion, we define the WiFi benefit as the improvement of its per-user throughput, as

$$z_k^w = (1 - \rho_k) R (N_k - n_k) - R (N_k) , \forall k \tag{3.35}$$

On the other hand, the LTE benefit can be defined as the amount of leftover unlicensed time slots for the existing LTE users, as

$$z^c = \sum_{k=1}^{K} \left(\rho_k - \rho_k^0 (n_k) \right) \tag{3.36}$$

where $\rho_k^0(nk)$ represents the minimum required unlicensed time slots in the k-th WiFi AP to keep the QoS of those n_k transferred users unaffected. In the above, both ρ_k and ρ_k^0 have been normalized according to the total number of unlicensed time slots.

From Eq. (3.36), larger LTE benefit can be achieved if more unlicensed time slots are allocated to LTE. However, this will decrease the WiFi benefit as observed from Eq. (3.35). Therefore, the values of $n_k, \forall k$ and $\rho_k, \forall k$ should be carefully selected to balance the benefits between LTE and WiFi. In the following, we will develop an algorithm to determine the optimal $n_k, \forall k$, and $\rho_k, \forall k$. Before that, we first calculate the minimum required unlicensed time slots, $\rho_k^0 (n_k)$, when transferring n_k users to LTE-U in the three different user transfer schemes , which is essential for analyzing the optimal values.

Fig. 3.8 Geometrical
relationship between L, d_i, r_i
and Θ_i

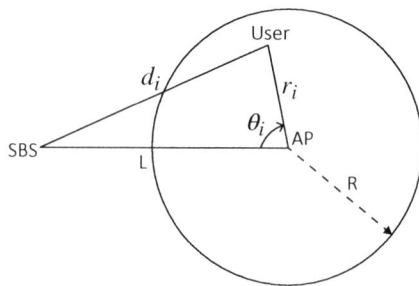

3.2.3 Minimum Required Unlicensed Resource

To calculate the minimum required unlicensed time slots, the ergodic capacity of
those transferred WiFi users should be first analyzed. Since each WiFi AP can be
independently considered, we omit subscript k and denote L as the distance between
the SBS and the WiFi AP for notation simplicity. Assume that $L > R$, where R is
the radius of the WiFi AP, as in Fig. 3.8.

3.2.3.1 Random Transfer

In this scheme, WiFi users are randomly transferred to the LTE-U system without
exploiting either geometrical information or CSI. As in Fig. 3.9, we denote the
distance and the angle between the WiFi AP and WiFi user i as r_i and θ_i,
respectively. The distance between WiFi user i and the SBS can be expressed as
$d_i = \sqrt{L^2 + r_i^2 - 2Lr_i \cos\theta_i}$.

We first derive the probability distribution function (PDF) of d_i. Assuming that
the WiFi users are uniformly distributed within the coverage of the WiFi AP, then
the PDF and the cumulative distribution function (CDF) of the distance between the
WiFi user and the WiFi AP, r_i, can be, respectively, expressed as

$$f_r(r_i) = \frac{2r_i}{R^2}, \; F_r(r_i) = \frac{r_i^2}{R^2}, 0 \le r_i \le R \tag{3.37}$$

Then, we can obtain the conditional CDF of d_i given θ_i as

$$F_{D|\theta}(d_i|\theta_i) = \Pr\left\{\sqrt{L^2 + r_i^2 - 2Lr_i \cos(\theta_i)} \le d_i\right\}$$

$$= F_r(L\cos\theta_i + x_i) - F_r(L\cos\theta_i - x_i) \tag{3.38}$$

where $x_i = \sqrt{(d_i^2 - (L\sin\theta_i)^2)}$. Taking the derivative of (3.38) with respect to d_i,
we get the conditional PDF of d_i given θ_i as

Fig. 3.9 The WiFi benefit under different L and N. (**a**) WiFi benefits versus the distance between SBS and WiFi AP, $N = 40$. (**b**) WiFi benefits versus the total number of WiFi users, $L = 16$ m

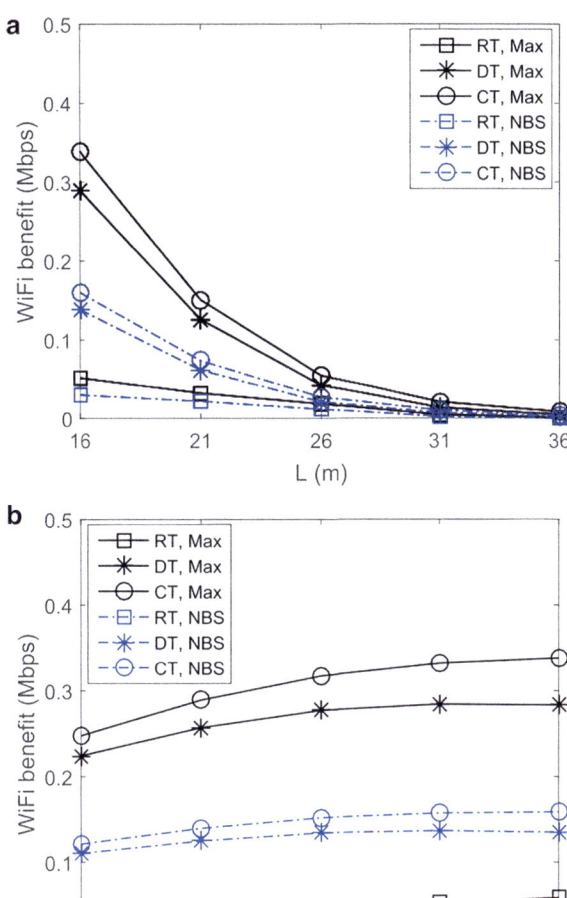

$$f_{D|\theta}\left(d_i | \theta_i\right) =$$

$$\begin{cases}
\dfrac{8d_i L \cos\theta_i}{R^2 x_i}, & |L\sin\theta_i| \leq d_i < L,\ 0 \leq \cos\theta_i < \cos\theta_H, \\[2ex]
\dfrac{8d_i L \cos\theta_i}{R^2 x_i}, & |L\sin\theta_i| \leq d_i < Y_i,\ \cos\theta_H \leq \cos\theta_i < \cos\theta_L, \\[2ex]
\dfrac{2d_i \left(L\cos\theta_i + x_i\right)}{R^2 x_i}, & L \leq d_i \leq Y_i,\ \cos\theta_i < \cos\theta_H, \\[2ex]
\dfrac{2d_i \left(L\cos\theta_i - x_i\right)}{R^2 x_i}, & Y_i \leq d_i \leq L,\ \cos\theta_i \geq \cos\theta_H, \\[2ex]
0, & \text{otherwise}
\end{cases}$$

where $\theta_H = \arccos\frac{R}{2L}$, $\theta_L = \arccos\frac{R}{L}$, and $Y_i = \sqrt{R^2 + L^2 - 2RL\cos\theta_i}$.
Furthermore, assuming that θ_i is uniformly distributed, the PDF of θ_i can be expressed as

$$f(\theta_i) = \frac{1}{2\pi}, \quad 0 \le \theta_i \le 2\pi \tag{3.39}$$

By using $f(\theta_i)$ and de-conditioning with respect to θ_i, we obtain the PDF of d_i as

$$f_D(d_i) = \frac{1}{2\pi} \int_0^{2\pi} f_{D|\theta}(d_i|\theta_i)\, d\theta_i \tag{3.40}$$

Then, the ergodic capacity of the i-th user with distance d_i can be expressed as

$$\mathbb{E}_H\left[b\log(1 + \gamma_i) \,|\, D = d_i \right] \tag{3.41}$$

where b is the bandwidth of a resource block and the expectation is over the fading fields. Furthermore, since

$$\Pr\left(\frac{P_i H_i d_i^{-\alpha_u}}{w} \ge t \right) = \Pr\left(H_i \ge tw P_i^{-1} d_i^{\alpha_u} \right) = e^{-tw P_i^{-1} d_i^{\alpha_u}} \tag{3.42}$$

and $\mathbb{E}_H[f(x)] = \int_0^\infty f'(x) \Pr(X \ge x)\, d_x$ if $f'(x)$ is a nonnegative and monotonically increasing function, we have

$$\begin{aligned}
&\mathbb{E}_H\left[b\log\left(1 + \frac{P_i H_i d_i^{-\alpha_u}}{w} \right) \,|\, D = d_i \right] \\
&= b \int_0^\infty \frac{1}{1+t} \Pr\left(\frac{P_i H_i d_i^{-\alpha_u}}{w} \ge t \right) dt \\
&= b \int_0^\infty \frac{1}{1+t} e^{-tw P_i^{-1} d_i^{\alpha_u}} \, dt
\end{aligned} \tag{3.43}$$

Then, de-conditioning with respect to $D = d_i$, we finally obtain the ergodic capacity of the i-th user in the RT scheme as

$$\mathbb{E}^{\mathrm{RT}}[R_i] = b \int_{L-R}^{L+R} \int_0^\infty \frac{1}{1+t} e^{-tw P_i^{-1} d_i^{\alpha_u}} f_D(d_i) \, dt\, dd_i \tag{3.44}$$

Based on the above ergodic capacity expression, we can then find the required amount of unlicensed resources if this user is transferred to the LTE-U system.

3.2.3.2   Distance-Based Transfer

In this scheme, the distance information will be exploited. That is, the WiFi AP transfers the first n users with the shortest distances, i.e., $\{d_1, d_2, \ldots, d_n\}$, to the SBS, where

$$d_1 < \ldots < d_n < \ldots < d_N$$

According to [37], the PDF of the i-th shortest distance between the SBS and the WiFi user can be written as

$$f_{D_i}^{DT}(d_i) = i \binom{N}{i} f_D(d_i) \left(1 - F_D(d_i)\right)^{N-i} F_D^{i-1}(d_i) \tag{3.45}$$

where $F_D(d_i)$ is the CDF of d_i. Similar to the RT scheme, we finally obtain the ergodic capacity of the i-th shortest distance WiFi user in the DT scheme as

$$\mathbb{E}^{DT}[R_i] = b \int_{L-R}^{L+R} \int_0^\infty \frac{1}{1+t} e^{-tw P_i^{-1} d_i^{\alpha u}} f_{D_i}^{DT}(d_i) \, dt \, dd_i \tag{3.46}$$

3.2.3.3   CSI-Based Transfer

In this scheme, users are selected based on the CSI. The first n users with the largest channel gain will be transferred, i.e., $\{g_1, g_2, \ldots, g_n\}$, where

$$g_1 > \ldots > g_n > \ldots > g_N \tag{3.47}$$

and $g_i = H_i d_i^{-\alpha u}$, $\forall i$. Then, the CDF of g_i can be expressed as

$$F_G(g_i) = \int_{L-R}^{L+R} f_D(d_i) \left(1 - e^{-g_i d_i^{\alpha u}}\right) dd_i \tag{3.48}$$

According to [37], the PDF of the i-th largest SNR among N WiFi users can be expressed as

$$f_{G_i}^{CT}(g_i)$$
$$= N \binom{N-1}{i-1} f_G(g_i) (F_G(g_i))^{N-i} (1 - F_G(g_i))^{i-1} \tag{3.49}$$

where $f_G(g_i)$ is the PDF of g_i.

Similar to the RT scheme, we can obtain the ergodic capacity of the i-th largest SNR WiFi user in the CT scheme as

$$\mathbb{E}^{\mathrm{CT}}\left[R_i\right] = b \int_0^\infty \log\left(1 + \frac{P_i g_i}{w}\right) f_{G_i}^{\mathrm{CT}}\left(g_i\right) \mathrm{d}g_i \qquad (3.50)$$

3.2.3.4 Minimum Required Unlicensed Resource

We have now obtained the ergodic capacity for those users transferred to the LTE-U system in the three different schemes. In this section, the minimum required time slots for the SBS to support the n transferred users, $\rho^0\left(n\right)$, will be calculated. Assuming those transferred users achieve the same throughput as in the WiFi network, we have

$$\rho^0\left(n\right) = \frac{1}{B^{\mathrm{T}}} \sum_{i=1}^n \frac{R\left(N\right)}{\mathbb{E}\left[R_i\right]} \qquad (3.51)$$

where $\mathbb{E}\left[R_i\right]$ is the ergodic capacity of the i-th user and could be $\mathbb{E}^{\mathrm{RT}}\left[R_i\right]$ in Eq. (3.44), $\mathbb{E}^{\mathrm{DT}}\left[R_i\right]$ in Eq. (3.46), or $\mathbb{E}^{\mathrm{CT}}\left[R_i\right]$ in Eq. (3.50) depending on which scheme is adopted, $R\left(N\right)$ is the per-user WiFi throughput expression in Eq. (3.33), and B^{T} is the unlicensed bandwidth.

3.2.4 Joint User Transfer and Unlicensed Resource Allocation in Different Cases

We now develop the win–win strategy to balance the benefits between the WiFi and LTE systems by jointly optimizing the number of transferred users and the time slot allocated to the LTE-U system. In this section, we first look into the single-AP case to find some insightful results and then investigate the multi-AP case.

3.2.4.1 Win–Win Strategy in Single-AP Case

We first investigate the maximal benefit that the WiFi network can achieve. From the previous analysis, the WiFi system achieves its maximal benefit when it transfers the most number of users to the LTE-U system but relinquishes the least time slots, i.e., $\rho^0\left(n\right)$. Therefore, the maximal benefit can be formulated as

$$z_0^w = \max_n \left\{\left(1 - \rho^0\left(n\right)\right) R(N - n) - R(N)\right\} \qquad (3.52)$$

Note that the WiFi system will allocate $\rho^0\left(n\right)$ time slots to the LTE-U system, which can only support the transferred users. Therefore, there is no benefit for the LTE system.

Similarly, LTE will obtain its maximal benefit when WiFi transfers fewer users but relinquishes more time slots to it. To ensure the fairness of the remaining WiFi users, we should keep their average throughput unchanged. Therefore, the problem to maximize the LTE benefit can be expressed as

$$z_0^c = \max_{\{n,\rho\}} \{\rho - \rho^0(n)\} \tag{3.53}$$

subject to

$$(1 - \rho) R(N - n) = R(N) \tag{3.54}$$

Also, there is no benefit for WiFi under this circumstance. The above equation can be rewritten as

$$z_0^c = \max_n \left(1 - \frac{R(N)}{R(N-n)} \right) - \rho^0(n) \tag{3.55}$$

which can be solved by exhaustive searching.

The above two situations, i.e., the maximal WiFi benefit and the maximal LTE benefit, are not win–win situations since there is no benefit for either WiFi or LTE. In this section, we will find a win–win strategy for both systems. Since the overall time slots on the unlicensed spectrum are constrained, it is impossible to maximize the WiFi and LTE benefits simultaneously. Therefore, we need to find an efficient user transfer and unlicensed time slot allocation scheme to balance the benefit between LTE and WiFi.

Recalling that the WiFi system will transfer n users and simultaneously relinquish ρ time slots to the LTE SBS, the general win–win problem to balance the WiFi and LTE benefits can be expressed as

$$\max_{\{n,\rho\}} \left\{ z^c, z^w \right\} \tag{3.56}$$

subject to

$$0 \le n \le N, \tag{3.56a}$$

$$\rho^0(n) \le \rho \le 1 \tag{3.56b}$$

It is a multi-objective problem. Therefore, we will use the NBS method [16, 38] to develop a fair win–win resource allocation scheme.

Definition 4 $(z_{\text{NBS}}^c, z_{\text{NBS}}^w)$ is a Nash bargaining solution if it solves the following problem:

$$(z_{\text{NBS}}^c, z_{\text{NBS}}^w) = \max_{\{n,\rho\}} z^c z^w \tag{3.57}$$

subject to Eq. (3.56a) and (3.56b).

Setting the first order derivative of Eq. (3.57) on ρ as zero, we can obtain the optimal time slot allocation for a given transferred user number, n, as

$$\rho = \max \left\{ \frac{1 + \rho^0(n)}{2} - \frac{R(N)}{2R(N-n)}, \rho^0(n) \right\} \tag{3.58}$$

Then, we introduce the following theorem.

Theorem 3 *Let n^w, n^c, n^{NBS} be the optimal numbers of transferred users to maximize the WiFi benefit, the LTE benefit, and the NBS objective in Eq. (3.57), respectively. Then, we have $n^c \leq n^{NBS} \leq n^w$.*

This theorem can be intuitively explained as follows. Since the WiFi benefit will be increased if more WiFi users are transferred, n^w should be the largest. On the contrary, n^c is the smallest since transferring less WiFi users would increase the LTE benefit. Meanwhile, the NBS strategy aims to make a fair balance between the two benefits. Therefore, n^{NBS} should be within $[n^c, n^w]$.

According to Theorem 1, we can obtain the optimal n^{NBS} by exhaustive searching from $[n^c, n^w]$, which is narrowed as compared to $[0, N]$. In this section, we briefly discuss the extension of the proposed algorithm into the downlink case. The WiFi downlink throughput expression is different from that of the uplink in Eq. (3.33) since downlink is contention free. Suppose the overall downlink saturation throughput is R^D. Then the average throughput for each user could be written as

$$R(n) = \frac{R^D}{n} \tag{3.59}$$

Similar to the above analysis, the WiFi benefit and the LTE benefit in the downlink case can be expressed as

$$z^w = (1 - \rho) \frac{R^D}{N - n} - \frac{R^D}{N} \tag{3.60}$$

and

$$z^c = \rho - \rho^0(n) \tag{3.61}$$

respectively.

We can easily demonstrate that if $\rho^0(n) < \frac{n}{N}$, for a given n, then both WiFi and LTE can achieve positive benefits simultaneously[1]. We can use the similar approach as in the uplink case to develop the win–win strategy for the downlink case, which is omitted due to page limit.

In this section, simulation results will be presented to evaluate the performance of the proposed win–win strategy. It is assumed that WiFi users are uniformly distributed within the coverage of the WiFi AP. We consider that the SBS and the WiFi APs have radii of 50 and 15 m, respectively. Both licensed and unlicensed spectra have bandwidth of 20 MHz and each LTE-U resource block has the bandwidth of 180 kHz. We adopt the IEEE 802.11n protocol working at 5 GHz and the parameters are listed in Table 3.2.

We first examine the performance in the single AP case. Figure 3.9a depicts the WiFi benefit with different distances between the SBS and the WiFi AP, i.e., L, in the three different user transfer schemes. When the WiFi AP is far away from the SBS, i.e., L is large, the channel gain between the WiFi users and the SBS would become small, leading to the decrease of WiFi benefit with L. Figure 3.9b shows the WiFi benefit with different numbers of WiFi users, i.e., N. Since the per-user WiFi throughput decreases with the user number, more WiFi users will be transferred to the LTE-U system as N goes large, leading to a large WiFi benefit. From both figures, the maximum WiFi benefit serves as the performance upper bound for any other user transfer schemes, e.g., the NBS-based strategy. Again, the CT scheme has the best performance, while the RT scheme performs the worst due to the reason indicated before.

Figure 3.10a, b shows the numbers of transferred users with different distances between the SBS and the WiFi AP, L, and different numbers of WiFi users, N, respectively. From the figures, the transferred user numbers for the maximum

Table 3.2 System parameters

Parameters	Settings
Noise power	-95 dBm
Path loss exp. α_u	5
Path loss constant C_u	0.16
Transmit power	30 dBm
$E[P]$	1500 byte
B^T	20 MHz
CW_{min}	16
CW_{max}	1024
R_{limit}	6
WiFi channel bit rate	130 Mbps
PHY header	192 bits
MAC header	224 bits
T_δ	20 μs
SIFS	16 μs
DIFS	50 μs
Slot time	9 μs
ACK	112 bits+PHY header
RTS	160 bits+PHY header
CTS	112 bits+PHY header

Fig. 3.10 Transferred user
number with different L and
N. (**a**) Transferred user
number with different
L, $N = 40$. (**b**) Transferred
user number with different
N, $L = 16$ m

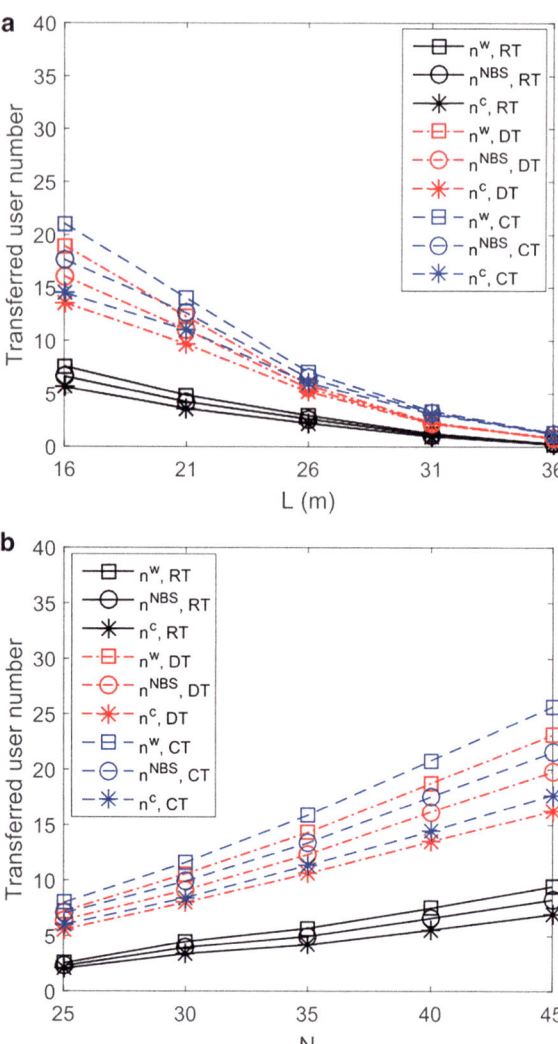

WiFi benefit and for the maximum LTE benefit represent the upper bound and the
lower bound, respectively, in all the three user transfer schemes, which confirms
Theorem 1. This can also be explained by the fact that the WiFi system intends to
transfer more users to maximize its own benefit, while the LTE system has exactly
opposite intention.

From Fig. 3.10a, as L goes large, fewer WiFi users would be transferred since
it would require more unlicensed time slots to serve them in the LTE-U system,
resulting in the low benefits for both systems. When L is large enough, no user
would be transferred to the LTE-U system since the benefit vanishes in this case.
From Fig. 3.10b, more users will be transferred to the LTE-U system as N increases

since it can significantly improve the WiFi performance and increase the overall benefits as well in this case. From both figures, the CT scheme can transfer the most number of WiFi users among all three schemes. The reason is that the WiFi users with high channel capacity will be transferred in this scheme and thus more users could be supported in the LTE-U system with the same amount of unlicensed resource than the other two schemes.

3.2.4.2   The Win–Win Strategy in Multiple-AP Case

We now investigate the joint user transfer and unlicensed resource allocation in the multiple-AP case. Similar to the single-AP case, we utilize the NBS as the optimization objective, which can be expressed as

$$\max_{\{n_k, \rho_k\}} z^c \prod_{k=1}^{K} z_k^w \tag{3.62}$$

subject to

$$0 \leq n_k \leq N_k, \forall k, \tag{3.62a}$$

$$\rho_k^0(n_k) \leq \rho_k \leq 1, \forall k \tag{3.62b}$$

It is quite challenging solving the problem in Eq. (3.62) since the two variables are nested as well as the optimal results are correlated among different APs.

In the following, we will develop an effective solution to solve it. First, due to the concavity and monotonicity of the $\log(\cdot)$ function, we can rewrite the objective function in Eq. (3.62) as

$$\max_{\{n_k, \rho_k\}} \log\left(z^c\right) + \sum_{k=1}^{K} \log\left(z_k^w\right) \tag{3.63}$$

Then, similar to the single-AP case, the following theorem expresses the optimal time slot allocation for a given user transfer result, i.e., $\{n_k\}, \forall k$.

Theorem 4 *The optimal ρ_k can be written as*

$$\rho_k = \max\left\{M - \frac{R(N_k)}{R(N_k - n_k)}, \rho_k^0(n_k)\right\}, \forall k \tag{3.64}$$

where $M = \frac{1}{1+k}\left(1 + \sum_{k=1}^{K}\left(\rho_k^0(n_k) + \frac{R(N_k)}{R(N_k - n_k)}\right)\right).$

Therefore, the LTE benefit and the WiFi benefit in Eqs. (3.35) and (3.36) can be rewritten as

$$z^c = \frac{K - \sum\limits_{k=1}^{K} \left(\rho_k^0 (n_k) + \frac{R(N_k)}{R(N_k - n_k)} \right)}{1 + K} \tag{3.65}$$

and

$$z_k^w = z^c R (N_k - n_k), \forall k \tag{3.66}$$

respectively, and the problem in Eq. (3.62) can be equally transformed into

$$\max_{\{n_k\}} (1 + K) \log \left(z^c \right) + \sum_{k=1}^{K} \log \left(R (N_k - n_k) \right) \tag{3.67}$$

subject to Eq. (3.62a).

Now the only remaining control variable is the number of transferred users, i.e., $\{n_k\}, \forall k$. Let n_k^{NBS} denote the optimal number of transferred users from WiFi AP k to maximize the NBS objective in Eq. (3.67). Then, we introduce the following properties.

Theorem 5 *Let N_k^w and N_k^c be the optimal numbers of the transferred users from WiFi AP k to maximize the WiFi benefit and the LTE benefit if only WiFi AP k is considered, respectively. That is $n_k^w = \arg\max\limits_{\{n_k\}} \left(1 - \rho_k^0 (n_k) \right)$*
$R (N_k - n_k) - R (N_k)$ and $n_k^c = \arg\max\limits_{\{n_k\}} \rho_k - \rho_k^0 (n_k)$. Then, we have $n_k^c \leq n_k^{\text{NBS}} \leq n_k^w, \forall k$.

Theorem 6 *Suppose $N_k = N, \forall k$, and $L_1 > L_2 > \ldots > L_k$, then we have $n_1^{\text{NBS}} \leq n_2^{\text{NBS}} \leq \ldots \leq n_K^{\text{NBS}}$.*
From Theorem 4, the WiFi APs near the SBS will transfer more users than those WiFi APs far away from the SBS. It is rather intuitive since fewer unlicensed time slots are required to serve those WiFi users in the nearby APs as compared to the far away APs. Therefore, transferring more users from nearby APs can generate larger benefits for both networks.

According to Theorem 3, we can develop an exhaustive search method to find the optimal n_K^{NBS}. The search space of WiFi AP k now becomes $[n_k^c, n_k^w]$, which is narrowed from $[0, N_k]$. However, it is still very complicated since a K-dimensional searching should be performed. In order to reduce the computational complexity, we develop a benchmark solution in the following. The idea of the benchmark solution is converting the K-dimensional searching into K 1-dimensional searching. That is, the $n_K^{\text{NBS}}, \forall k$ is determined by the algorithm developed in the single-AP case as if there were only WiFi AP k in the system. In the following section, we will compare the performance between the optimal algorithm and the benchmark algorithm.

In the multi-AP case, we consider one SBS and four WiFi APs that have distances of 31, 26, 21, and 16 m, to the SBS. The number of users in each WiFi AP is 46. Since DT is the most relevant scheme from a practical implementation perspective, we only consider the DT scheme in the sequel.

Table 3.3 presents the transferred user numbers and the amount of allocated unlicensed time slots in each WiFi AP for the maximum WiFi benefit, the maximum LTE benefit, the NBS strategy, and the benchmark strategy, respectively. From the table, more users will be transferred from the AP nearby (AP 4) than the AP far away (AP 1), which confirms Theorem 4. On the other hand, the amount of unlicensed time slots relinquished to the LTE SBS, i.e., ρ_k, increases with the transferred user number, n_k. This is rather intuitive since more unlicensed resources would be required if more WiFi users are transferred, as can be observed from Theorem 2. Moreover, the optimal number of transferred users in the NBS strategy is no smaller than the one that maximizes the LTE benefit, while it is no larger than the one that maximizes the WiFi benefit, which confirms Theorem 3.

In Fig. 3.11, we show the objective function in Eq. (3.62) for the NBS and the benchmark strategies where all WiFi APs have the same user number, N. From the figure, we can conclude that the NBS strategy can provide better fairness for WiFi and LTE coexistence than the benchmark strategy, whereas the latter bears lower computational complexity.

Table 3.3 Transferred user number and relinquished time slot number of each AP in the DT scheme

| Algorithm | (n_k, ρ_k) | | | |
	(n_1, ρ_1)	(n_2, ρ_2)	(n_3, ρ_3)	(n_4, ρ_4)
Max WiFi	(5,0.1305)	(9,0.2192)	(19,0.4462)	(27,0.5950)
NBS	(5,0.1305)	(9,0.2192)	(18,0.4303)	(23,0.5485)
Benchmark	(2,0.0544)	(9,0.2298)	(12,0.2912)	(23,0.5364)
Max LTE	(2,0.0583)	(9,0.2409)	(12,0.3167)	(19,0.4873)

Fig. 3.11 NBS objective function comparison

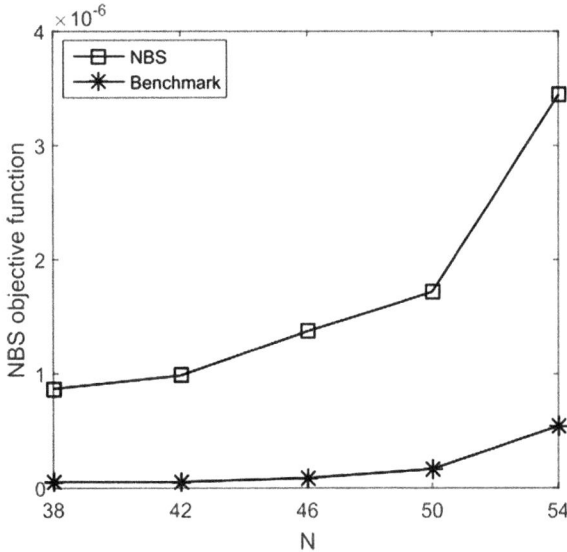

References

1. Aijaz, A., Aghvami, H., Amanim, M.: A survey on mobile data offloading: technical and business perspectives. IEEE Wirel. Commun. **20**(2), 104–112 (2013)
2. Lee, K., Lee, J., Yi, Y., Rhee, I., Chong, S.: Mobile data offloading: How much can WiFi deliver? IEEE/ACM Trans. Netw. **21**(2), 536–550 (2013)
3. Bennis, M., Simsek, M., Czylwik, A., Saad, W., Valentin, S., Debbah, M.: When cellular meets WiFi in wireless small cell networks? IEEE Commun. Mag. **51**(6), 44–50 (2013)
4. Xin, K., Chia, Y.K., Sun, S., Chong, H.F.: Mobile data offloading through a third-party WiFi access point: an operator's perspective. IEEE Trans. Wirel. Commun. **13**(10), 5340–5351 (2014)
5. Dong, H., Wang, P., Niyato, D.: A dynamic offloading algorithm for mobile computing. IEEE Trans. Wirel. Commun. **11**(6), 1991–1995 (2012)
6. Jung, B.H., Song, N., Sung, D.K.: A network-assisted user-centric WiFi-offloading model for maximizing per-user throughput in a heterogeneous network. IEEE Trans. Veh. Technol. **63**(4), 1940–1945 (2014)
7. Singh, S., Dhillon, H.S., Andrews, J.G.: Offloading in heterogeneous networks: modeling, analysis, design insights. IEEE Trans. Wirel. Commun. **12**(5), 2484–2497 (2013)
8. Ye, Q., et al.: User association for load balancing in heterogeneous cellular networks. IEEE Trans. Wirel. Commun. **12**(6), 2706–2716 (2013)
9. Yun, S., Yi, Y., Cho, D., Mo, J.: The economic effects of sharing femtocells. IEEE J. Sel. Areas Commun. **30**(3), 595–606 (2012)
10. Lin, G., Oosifidis, G., Huang, J., Tassiulas, L., Li, D.: Bargaining-based mobile data offloading. IEEE J. Sel. Areas Commun. **32**(6), 1114–1125 (2014)
11. Qualcomm: Qualcomm research LTE in unlicensed spectrum: harmonious coexistence with Wi-Fi [Online] (June 2014). Available: https://www.qualcomm.com/media/documents/files/lte-unlicensedcoexistence-whitepaper.pdf
12. Babaei, A., Andreoli-Fang, J., Hamzeh, B.: On the impact of LTEU on Wi-Fi performance. In: Proceedings of the IEEE 25th Annual International Symposium on Personal, Indoor, and Mobile Radio Communications (PIMRC), Washington, DC, USA, Sept, pp. 1621–1625 (2014)
13. Elsherif, A.R., Chen, W.P., Ito, A., Ding, Z.: Resource allocation and inter-cell interference management for dual-access small cells. IEEE J. Sel. Areas Commun. **33**(6), 1082–1096 (2015)
14. Bianchi, G., Tinnirello, I.: Kalman filter estimation of the number of competing terminals in an IEEE 802.11 network. In: Proceedings of the INFOCOM, San Francisco, CA, USA, Apr, pp. 844–852 (2003)
15. Bianchi, G.: Performance analysis of IEEE 802.11 distributed coordination function. IEEE J. Sel. Areas Commun. **18**(3), 535–547 (2000)
16. Malone, D., Duffy, K., Leit, D.: Modeling the 802.11 distributed coordination function in nonsaturated heterogeneous conditions. IEEE/ACM Trans. **15**(1), 159–172 (2007)
17. Ekici, O., Yongacoglu, A.: IEEE 802.11a throughput performance with hidden nodes. IEEE Commun. Lett. **12**(6), 465–467 (2008)
18. Andrews, J.G., Baccelli, F., Ganti, R.K.: A tractable approach to coverage and rate in cellular networks. IEEE Trans. Commun. **59**(11), 3122–3134 (2011)
19. Dhillon, H.S., Ganti, R.K., Baccelli, F., Andrews, J.G.: Modeling and analysis of K-tier downlink heterogeneous cellular networks. IEEE J. Sel. Areas Commun. **30**(3), 550–560 (2012)
20. Geisser, S. Johnson, W.M.: Modes of Parametric Statistical Inference. Wiley, Hoboken (2006)
21. Bleicher, A.: A surge in small cell sites. IEEE Spectr. **50**(1), 38–39 (2013)
22. 3GPP: Study on licensed-assisted access to unlicensed spectrum (Release 13). Technical Report. TR 36.889, version 13.0.0 (June 2015)
23. Zhang, R., Wang, M., Cai, L.X., Zheng, Z., Shen, X., Xie, L.: LTE Unlicensed: the future of spectrum aggregation for cellular networks. IEEE Wirel. Commun. **22**(3), 150–159 (2015)

24. Zhang, H., Chu, X., Guo, W., Wang, S.: Coexistence of Wi-Fi and heterogeneous small cell networks sharing unlicensed spectrum. IEEE Commun. Mag. **53**(3), 158–164 (2015)
25. Wang, S., Guo, W., Farrell, T.O.: Energy efficiency evaluation of SISO and MIMO between LTE-femtocells and 802.11n networks. In: Proceedings of the IEEE Vehicular Technology Conference (VTC Spring), pp. 1–5 (2012)
26. Harmonized European Standard, Broadband Radio Access Networks (BRAN), 5 GHz High Performance RLAN, ETSI EN 301 893 (June 2012)
27. Nihtila, T., et al.: System performance of LTE and IEEE 802.11 coexisting on a shared frequency band. In: Proceedings of the IEEE Wireless Communications and Networking Conference (WCNC), Shanghai, China, pp. 1038–1043 (2013)
28. Almeida, E., et al.: Enabling LTE/WiFi coexistence by LTE blank subframe allocation. In: Proceedings of IEEE International Conference on Communications (ICC), Budapest, Hungary, June, pp. 5083–5088 (2013)
29. Sagari, S., Baysting, S., Saha, D., Seskar, I., Trappe, W., Raychaudhuri, D.: Coordinated dynamic spectrum management of LTEU and Wi-Fi networks. In: Proceedings of the IEEE International Symposium on Dynamic Spectrum Access Networks (DySPAN), pp. 209–220 (2015)
30. Liu, F., Bala, E., Erkip, E., Beluri, M., Yang, R.: Small cell traffic balancing over licensed and unlicensed bands. IEEE Trans. Veh. Technol. **64**(12), 5850–5865 (2015)
31. Chen, J.Q., Yu, G., Yin, R., Maaref, A., Li, G.Y., Huang, A.: Energy efficiency optimization in licensed-assisted access. IEEE J. Sel. Areas Commun. **34**(4), 723–734 (2016)
32. Chen, Q., Yu, G., Shan, H., Maaref, A., Li, G.Y., Huang, A.: Cellular meets WiFi: traffic offloading or resource sharing? IEEE Trans. Wirel. Commun. **15**(5), 3354–3367 (2016)
33. Ismail, M., Zhuang, W.: A distributed multi-service resource allocation algorithm in heterogeneous wireless access medium. IEEE J. Sel. Areas Commun. **30**(2), 425–432 (2012)
34. Kim, S., Kim, B., Fang, Y.: Downlink and uplink resource allocation in IEEE 802.11 wireless LANs. IEEE Trans. Veh. Technol. **54**(1), 320–327 (2005)
35. Blaszczyszyn, B., Karray, M.K., Keeler, H.P.: Using Poisson processes to model lattice cellular networks. In: Proceedings of the IEEE International Conference on Computer Communications (INFOCOM), pp.773–781 (2013)
36. Lin, X., Andrews, J.G., Ghosh, A.: Modeling, analysis and design for carrier aggregation in heterogeneous cellular networks. IEEE Trans. Commun. **61**(9), 4002–4015 (2013)
37. Balakrishnan, N., Cohen, A.C.: Order Statistics and Inference: Estimation Methods, Statistical Modeling and Decision Science. Academic, New York (1991)
38. Owen, G.: Game Theory, 3rd edn. Academic, New York (2001)

Chapter 4
Mobile Data Application in Mobile Network

Abstract This chapter focuses on mobile networks: mobile data analysis and application technology in mobile networks. In recent years, many new network architectures and technologies have emerged in mobile networks, such as D2D networks, green communication networks, and sensing networks. This chapter will be elaborated in turn: mobile data application technology in D2D network, mobile data application technology based on energy saving in green communication network, and mobile application technology in perceived network.

Keywords Mobile networks · Network architectures · D2D networks · Green communication networks · Sensing networks · Energy saving

4.1 Mobile Data Application in D2D Networks

4.1.1 D2D Networks: Concepts and Applications

According to Cisco's mobile network outlook report [1], global mobile data traffic will grow to 292EB in 2019, of which 97% data traffic will be generated by smart mobile devices. One interesting phenomenon is that content requests are highly concentrated, so some popular contents are repeatedly requested. The redundant content request will burden the network backhaul a lot. Traditional centralized mobile cloud computing services cannot effectively match the explosive growth of the massive network edge data. A new paradigm called mobile D2D networks is proposed in mobile computing. ETSI first proposed the concept of mobile D2D networks in 2014. D2D refers to a technology that deeply integrates base stations and Internet services based on the 5G evolution architecture. Base station (BS) with cache ability and even mobile user with the exact content can provide flexible content offloading. Content offloading exploits complementary networks to deliver content data interested, thereby reducing network backhaul overload. Current mobile offloading techniques mainly involve D2D server by using WiFi, femtocell, etc. Due to the limited number and coverage of the D2D server, the content offloading with great dynamics is also limited.

© Springer Nature Switzerland AG 2019

H. Jiang et al., *Mobile Data Mining and Applications*, Information Fusion and Data Science, https://doi.org/10.1007/978-3-030-16503-1_4

In this chapter, we first propose the content offloading dynamic decision scheme based on the prediction of opportunistic mobile content providers. The prediction aims to find the possible content providers with the popular content users may request. During the dynamic content offloading decision, we consider the mobile content consumers and content providers including mobile users and D2D servers (e.g., BS) as the relationship of "leader–followers," i.e., Stackelberg game. In realistic network situations, the system state information is usually imperfect. For this scenario, we then propose the online content offloading scheme to handle the imperfect system state information. In the scheme, we propose to learn from historical records and make online decisions through gradient bandit.

The main contributions of the chapter include:

* Considering the characteristics that content requests are highly concentrated and some of the popular contents are repeatedly requested, we propose an opportunistic content offloading scheme by predicting the opportunistic content providers with popular and repeatedly cached contents among mobile users and edges for D2D paradigm.
* To handle the dynamic network situations, we propose a content offloading scheme among BS and mobile users modeled by Stackelberg game. During the dynamic offloading decision process, we utilize the predictions of opportunistic mobile content providers. In this case, the offloading decision with Stackelberg game can start from a reasonable initial pricing that speeds the convergence to equilibrium.
* In realistic network situations, the system state information is imperfect especially in dynamic network situations. we propose an online content offloading scheme by using gradient bandit, in order to maximize the system utility.
* Our evaluations are based on the real dataset from China Mobile Communications Corporation. Our dynamic content offloading decision scheme provides flexible and low-cost offloading with fast convergence. Our online content offloading scheme is with good efficiency in terms of content load.

The rest of this chapter is organized as follows: system model and overview of our method are presented in Sect. 4.2. Section 4.3 proposes a prediction algorithm for opportunistic content providers.

Content offloading for mobile D2D networks systems has attracted significant attention in recent years. Joseleal [2] presented a lightweight and efficient framework called user-level online offloading framework (ULOOF), for mobile computation offloading. ULOOF is equipped with a decision engine that minimizes remote execution overhead, while not requiring neither superuser privileges on the mobile device nor modifications to the underlying operating system. Tang and He [3] study the multi-user computation offloading problem in D2D from a behavioral perspective. Based on the framework of prospect theory (PT), they cast the users' decision-making of whether to offload or not as a PT-based non-cooperative game and propose a distributed computation offloading algorithm to achieve the Nash equilibrium. The result shows under the PT model, which considers the irrational and subjective user behavior, the number of offloading users becomes smaller than the classical EUT model. Ketyk et al. [4] presented an overview of NP-hard problems and methods related to deployment, resource sharing, load balancing, and fairness among multiple users in 5G mobile networks. The simulation results demonstrate that heuristic approaches are needed as the exact solution (Mulknap) is unable to get a result within a short running time. Chen et al. [5] propose a game theoretic approach for the computation offloading decision- making problem among multiple mobile device users for mobile edge cloud computing. The proposed approach can achieve efficient computation offloading performance.

Furthermore, for the aspect of current mobile data content offloading methods, current mobile offloading techniques mainly involve D2D server by using WiFi [6], femtocell [7], etc.

WiFi is a wireless access technology based on the IEEE 802.11 series of standards. At present, most smart devices are equipped with WiFi network interfaces and many places have WiFi network coverage. Moreover, WiFi's cost is lower than the 3G, 4G's, so it has great advantages to offload mobile data content using already deployed WiFi network. Lee et al. [8] analyzed the performance of the WiFi-based 3G mobile data offloading scheme by analyzing the mobile devices of 100 volunteers accessing the WiFi data in 2 weeks and a half. The results show that WiFi can effectively offload about 65% of mobile data while saving 55% of power consumption. Balasubramanian et al. [9] studied the data of mobile vehicles accessing 3G and WiFi networks in three cities. The availability of 3G and WiFi networks was 87% and 11%, respectively. Due to the small coverage area of WiFi network, the offloaded data content is limited. WiFi-based content offload was implemented through delay transmission and fast handover. First, predict the availability of WiFi. If the transmission can be completed within the tolerable delay time, the data transmission is delayed. For delay-sensitive data, it is quickly switched to the 3G network. The results show that when the tolerated delay time is 60 s, about 45% of the data volume can be offloaded. However, using WiFi to offload content has limitations: it is mainly used as a bandwidth access for indoor and has a limited coverage.

Han demonstrated the feasibility of mobile data content offloading based on opportunistic communications and analyzed the method of random, heuristic, and greedy algorithms [10]. Izumikawa combined cellular network with opportunistic network and utilized a storage-carry-forward mechanism to offload uplink content [11]. Hui further enhances offloading efficiency by identifying social network and transmitting specific content to a social group [12]. Han exploited heuristic algorithm to select k users as the initial set, and these users further forward the content to other users through short-range wireless connections such as Bluetooth and WiFi [13]. Li et al. [14] and Andreev et al. [15] exploited DTN networks (delay tolerant networks) for mobile data offloading.

The femtocell is a small, low-cost, low-power base station that is typically deployed at home and at work [16]. Compared with WiFi, which mainly focuses on unlicensed frequency bands, the femtocell mainly focuses on the licensed frequency band. The content generated by the user's mobile device is not transmitted via the macrocell network base station. Therefore, the femtocell can effectively reduce the load pressure on the cellular network and improve the service quality of users. Ramaswamy and Das [17] considered the interference between cells as a Gaussian random variable and explored the femtocell's backlinking capability. The literature [18] studied femtocell mobility management and access control, using access control to improve interference. To control offloading and to achieve the required balance of users and traffic served by each network tier, in [19] the authors quantify offloading and discuss different techniques that can be used to offload users from the macro access network to the femto access network. Three offloading techniques, namely offloading via power control, offloading via biasing, and offloading via increasing the relative intensity of FAPs, have been investigated. To achieve secure data transmission, in [20], femto-cloud architecture is used where the application is offloaded from the mobile device to the cloud through the secure and low-power base station femtocell. Simulation results present that using femto-cloud architecture 70–83% and 52–66% power savings are achieved than using macrocell and microcell base stations, respectively, while offloading an application to the cloud.

However, most of the proposed content offloading frameworks in D2D focuses on mobile data content offload in fixed scenarios. They do not consider user mobility and cannot predict when users can be in proximity to provide content. Also, these studies need to rely on specific known content offload requirements [7, 21–23], thereby being unable to make optimal offloading decisions based on the content load trends in dynamic scenarios. Then, based on this, we design a dynamic content offloading decision scheme, which performs content offloading by using moving mobile user's opportunity content provided in dynamic networks.

4.1.2   Dynamic Content Offloading in D2D Networks

We consider opportunistic content offloading scenario for D2D paradigm as shown in Fig. 4.1. Each base station is equipped with a D2D server and the mobile data provided by the BS is stored in the D2D server. For content consumer, content can be obtained either from the D2D server (e.g., BS) or from other mobile users caching the same content. For example, the UE1 cache the popular content that may be also needed by UE2. UE2 can obtain the content either from the D2D server or from user1 that depends on the cost. Our scheme provides the prediction ability to find the possible content providers and the opportunistic content offloading by fully utilizing the cache ability of mobile users.

We formulate the problem in two steps: In the first step, we make predictions for opportunistic content providers. In the second step, we first propose the content offloading dynamic decision scheme based on the prediction results in the first step. Then, in realistic dynamic network situations where the information is imperfect, we propose an online content offloading scheme based on the reinforcement learning.

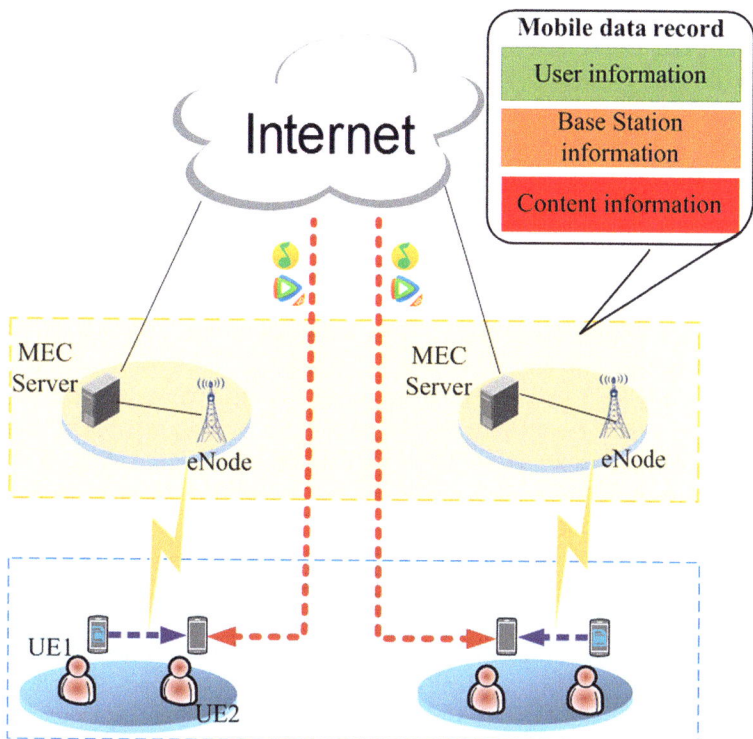

Fig. 4.1 System architecture

In order to perform content offloading for each content consumer, we need to predict which users can provide content for him. We consider a 1-h time scale, and the day is divided into 24 h. In each time period, we will first seek for the opportunistic content provider for each content consumer according to the rules in Eq. (4.1). The content providers and the content consumer should have an intersection in time and space.

$$
\begin{cases}
\text{Loc}_i = \text{Loc}_j \\
(\text{St}_i, \text{Et}_i) \cap (\text{St}_j, \text{Et}_j) \neq \emptyset
\end{cases}
\tag{4.1}
$$

where $\text{Loc}_i, \text{Loc}_j$ are the users' locations in geographic space denoted by the location information of base station visited by user. For user i and j, the start and end times of a data stream are denoted by $(\text{St}_i, \text{Et}_i)$ and $(\text{St}_j, \text{Et}_j)$, respectively.

Based on the definition of user opportunistic content provide, we consider a user opportunistic content provide network, which consists of users denoted by nodes and the opportunistic content provide between users denoted by edges.

Given an undirected graph $G = (V, E)$, where V is a set of smartphone users, and for a pair of users $x \in U$ and $y \in V$, $(x, y) \in E$, if the information of this two users satisfy the conditions in Eq. (4.1). The user opportunistic content provide prediction problem is transformed into the problem of link prediction in the graph.

After the link prediction, for each content consumer, we can predict other users who are in the same base station and have the intersection of the Internet data flow time. Further, among these users, we need to select mobile users who have an intersection with the content consumer user on the interest content as content providers. Specifically, for a single user u_i, $E\left(u_i\right) = \left\{v_1, v_2, \ldots, v_{m(u_i)}\right\}$ presents the set of predicted users who have a link with u_i in the graph during the t period in the future day. Meanwhile, the set of user interest content is recorded as $I\left(u_i\right)$, $I\left(u_i\right) \cap \left(v_j\right) \neq \emptyset$ represents that there is an intersection between the two users' interest content collections and user v_j can serve as a content provider and provide content to content consumer.

After filtering by the content dimension, for a content consumer, the set of his content providers is recorded as $\phi\left(u_i\right) = \left\{v_1, v_2, \ldots, v_{n(u_i)}\right\}$, where $n\left(u_i\right)$ is the number of content providers. Then, we develop content offloading schemes in dynamic network.

After determining the content providers for each content consumer, we need to make a decision for the content consumer, that is, whether to get content from the BS or get from the content providers. The content that the consumers obtain from other users is the offloading content.

In the dynamic network, we formulate the content offloading problem as a Stackelberg game in which content demander is the leader, and BS and content providers are the followers. As shown in Fig. 4.1, the content consumer's price strategy is recorded as p, p_0 represents the initial pricing stated by the content consumers. It can be obtained by predicting the user's consume content for a specific time period in the future day, as Eq. (4.2):

$$p_0 = Q_t * \min\left(c_i\right) \tag{4.2}$$

where Q_t denotes the predicted content of a single user, and c_i denotes the unit content cost of the BS and the content providers.

The content policy combination of n followers is recorded as $f = \{f_1, f_2, \ldots, f_n\}$, the content policy set is recorded as F. The content consumer's utility function is recorded as $U_0\left(p, f\right)$, the utility function for follower i is recorded as $U_i\left(p, f\right)$, $i \in \{1, 2, \ldots n\}$ and $p \in P, f \in F$. Given a content consumer's price strategy p, the follower's Nash equilibrium point set of non-cooperative game is recorded as $N\left(p\right) = \left\{f^* = \left(f_1^*, f_2^*, \ldots, f_n^*\right)\right\}$. If any strategy satisfies Eq. (4.2), then f^* is a Nash equilibrium point with parameters for the non-cooperative game. In the two-stage single-master multi-slave Stackelberg game, if there is any combination of strategies to $\left(p^*, f^*\right) \in P \times F$, $f^* \in N\left(p\right)$ satisfy the formula (4.3), then $\left(p^*, f^*\right)$ is called an equilibrium point of the game. That is, the subgame perfect Nash equilibrium.

$$U_i\left(p, f_i^*, f_{-i}^*\right) \geq U_i\left(p, f_i, f_{-i}^*\right) \tag{4.3}$$

$$U_0\left(p^*, f^*\right) \geq U_0\left(p, f^*\right) \tag{4.4}$$

Nash equilibrium is the solution to the problem of non-cooperative game. The core goal of establishing Stackelberg game in this step is to find solutions to the content offloading problem. We need to get the pricing p when all game participants reach the Nash equilibrium. Then, we can obtain the best content policy that follower i provides to content consumers. In this scenario, we are looking for the pricing point that maximizes content consumer's utility, as shown in Eq. (4.5).

$$p = \mathrm{argmax}_q u_0, \text{subjectto} p \geq p_0 \tag{4.5}$$

In realistic dynamic network, the system state information is imperfect. The user data content consume in the next day is unknown, but the mobile data content offloading scheme should be made beforehand. We design utility functions for the BS, the content provider, and the content consumers.

For the BS, the utility function is composed of the benefits and costs of providing content to the content consumers. Assume the total user content consume is q, the possible offload content is f, and the offload ratio is x, the BS's real content load is $q-fx$. The unit price $p = a \log\left(1 + b \cdot q\right)$ is the dynamic pricing based on the content, where a, b are the parameters that map the content to the price range [8, 24]. When it is in the peak period of content, it restricts the use of data content by increasing the price. When it is in the trough period of content, it encourages users to use data content by appropriately lowering the price. The unit cost of data content is e, and the BS's utility function is as follows:

$$U_1\left(x\right) = \left(p - e\right) \cdot \left(q - fx\right) \tag{4.6}$$

The content provider gains revenue by providing content data to users. The total amount of provided data is the product of the possible offloading content f and the offloading ratio x. Since data storage occupies the limited space of the terminal and data transmit consumes the CPU and power, the unit price and cost of content provider are denoted as d and g. The utility function of the content provider is as follows:

$$U_2\left(x\right) = \left(d - g\right) \cdot fx \tag{4.7}$$

For the content consumer, the utility function is composed of data content income and payment expenses, including the payment for BS and content provider.

$$U_3\left(x\right) = \alpha \log\left(1 + q\right) - p\left(q - fx\right) - dfx \tag{4.8}$$

where the first item represents the benefit when the content consumer obtains the total content q, α is a parameter related to the user experience, and the logarithm function is taken [25, 26]. The second item is the fee paid to the BS, where p is the unit content pricing, and $q - fx$ is the actual content load of the BS. The third item is the fee paid to the content provider, where d represents the unit content pricing of the content provider, and fx is the total content provided by the content provider.

The goal of mobile data content offloading is to determine offloading scheme which maximizes the system utility function, where λ_1, λ_2, and λ_3 are normalized weights.

$$U(x) = \lambda_1 \cdot U_1(x) + \lambda_2 \cdot U_2(x) + \lambda_3 \cdot U_3(x)$$
$$\text{s.t. } \lambda_1 + \lambda_2 + \lambda_3 = 1$$
(4.9)

We consider 1-h time scale, and the day is divided into 24 h. In each time period, we will first seek for the opportunistic content provider for each content consumer. We construct complex network in which nodes represent users and edges represent the content provide relationship between users. By constructing a complex network, the user opportunistic content provider prediction problem is transformed into the problem of link prediction in the graph.

We analyzed multi-dimensional features from the three aspects of network topology, user mobility, and Internet behavior characteristics. Traditional topology features extracted from network can be exploited to measure the similarity of users. Three topology features including common neighbor index, Salton index, and Adamic–Adar index were selected [27]. The basic idea of the common neighbor index is that if the number of common neighbors of u and v is more, the possibility that there is an edge connecting these nodes is greater. For node u, the neighbor node set of u is $\Gamma(u)$ and degree of u is $k(u)$.

$$\text{CN}(u, v) = |\Gamma(u) \cap \Gamma(v)|$$
(4.10)

Salton index is also called cosine similarity. It further considers the effect of the product of two node degrees on the similarity.

$$\text{Salton}(u, v) = \frac{|\Gamma(u) \cap \Gamma(v)|}{\sqrt{k(u) \times k(v)}}$$
(4.11)

In the Adamic–Adar index, the smaller the degree of the common neighbor node, the greater its effect on the similarity. Therefore, the Adamic–Adar index takes the reciprocal of the node degree in the logarithm format.

$$\text{AA}(u, v) = \sum_{z \in \Gamma(u) \cap \Gamma(v)} \frac{1}{\lg k(z)}$$
(4.12)

In terms of user mobility features, the distance between the hotspot is exploited to characterize the similarity of users. The hotspot is the location with the highest access frequency. We use Haversine formula to calculate the distance in geographic space. Furthermore, the Pearson similarity and Kullback–Leibler divergence are commonly used indicators in the research of human mobility [28, 29]. Assume that the frequency sequences of user u and v accessing the base station are $u = \langle k_1, k_2, \ldots, k_i, \ldots, k_n \rangle$ and $v = \langle l_1, l_2, \ldots, l_i, \ldots, l_n \rangle$, where N is the total number of access base stations. It is calculated as follows:

$$\text{Pearson} = \frac{\sum_i \left(k_i - \bar{k}\right)\left(l_i - \bar{l}\right)}{\sqrt{\sum_i \left(k_i - \bar{k}\right)^2 \left(l_i - \bar{l}\right)^2}} \tag{4.13}$$

$$\text{KLD}\left(K \| L\right) = \sum_i k_i \lg \frac{k_i}{l_i} \tag{4.14}$$

Based on the classic indicator common neighbor index, Ref. [30] takes the weight information of the edges into account to further improve the prediction performances. Inspired by this, we take the number of times the content provider provide content as weights to obtain the similarity indicators based on the content provide behavior. In the following equation, $e(u, v)$ is the number of provide contents.

$$\text{Pro\_WCN}(u, v) = \sum_{z \in \Gamma(u) \cap \Gamma(v)} e(u, z) + e(z, v) \tag{4.15}$$

In terms of user Internet behavior features, we similarly introduce the features based on the duration of surfing the Internet and data content.

$$\text{Dur\_WCN}(u, v) = \sum_{z \in \Gamma(u) \cap \Gamma(v)} d(u, z) + d(z, v) \tag{4.16}$$

Moreover, we exploit the Jaccard similarity of Internet content. Assume the content category sets of user u and v are C_u and C_v, the Jaccard similarity is calculated as follows:

$$\text{Jaccard} = \frac{|C_u \cap C_v|}{|C_u \cup C_v|} \tag{4.17}$$

Finally, using these extracted features from the three aspects of network topology, user mobility, and Internet behavior as input information, random forest model is exploited to predict user content provide behavior.

In the dynamic network situation, the user content consumes and content provided are known by analyzing content consumes and predicting content providers, respectively. This chapter proposed a dynamic content offloading algorithm based on Stackelberg game.

From the analysis in the user content consume could be predicted by analyzing the known user history content consume information. Meanwhile, the user opportunistic content provide could be predicted based on the proposed prediction algorithm. Once the content provide relationship has been established among users, the content consumer could download interesting content from BS or content providers. However, the content offloading scheme based on opportunistic content providers does not always guarantee that consumers and providers adopt it in practice. In this section, we propose a mobile data offloading algorithm based on Stackelberg game as follows.

As shown in Fig. 4.2, in the Stackelberg game, there are two kinds of rational players: leaders and followers. The content consumers are considered as leaders who decide on the initial pricing parameters p. The BS and content providers are considered as followers. Each follower is a price-taker and offer the leader content to maximize his own utility. The part of data content that the leader obtains from the content providers is the offloaded content based on user opportunistic content provide. Our analysis and numerical results are carried out based on the equilibrium of this game.

We formulate the interactions between content consumer and content providers as a two-stage non-cooperative game. In the first stage, every user who asks for content proposes a price to every provider within its coverage area. In the second stage, every provider indicates the content volume it is willing to provide for user who has proposed to it. The content obtained from other users can be offloaded from cellular network. We then designed utility functions for the BS, the content providers, and content consumer, respectively. Specifically, we assume that user i has j followers including BS and $j - 1$ content providers, which could be predicted from the user opportunistic content provide model. Then, the utility of leader i is denoted as U_0, which we will describe later and rely on content as well as the unit cost of content. Similarly, the utility of every follower j is denoted as U_j. We prove the existence of Nash equilibrium in the constructed Stackelberg game model. Then, a distributed iterative algorithm is exploited to solve the optimal pricing and content offloading algorithm. Meanwhile, according to the average content consumption of

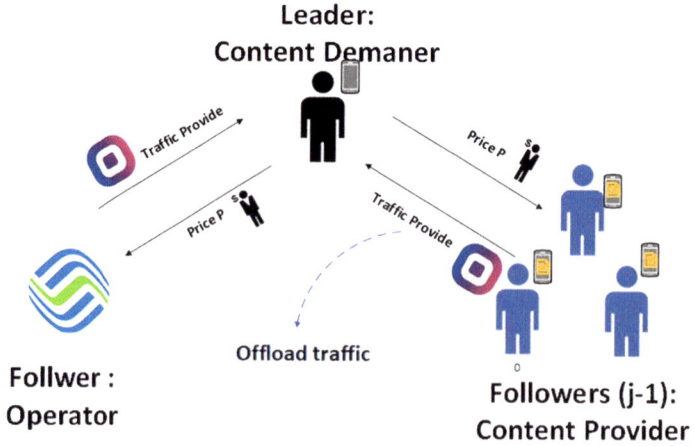

Fig. 4.2 Stackelberg game model

several days ago in different period, the total content consume of users in the future days can be predicted, which can be used for optimal solution.

We first design the utility function for the players in the Stackelberg game proposed in this chapter, and then we prove the existence of the Nash equilibrium in the game.

We designed utility functions for BS, content providers, and content consumer, respectively, to reflect the reward that participants receive. For the leader, the utility function of content consumer user is composed of data content income and payment expenses. $\sum_{j \in \text{Followers}} f_j$ is the amount of data content that followers provide for the leader. We took the form of a logarithm function to represent the leaders' gain for data content. The leader's utility function is calculated by

$$U_0 = \alpha \log \left(1 + \sum_{j \in \text{Followers}} f_j \right) - p \tag{4.18}$$

where α is a parameter related to user experience, and p denotes the total price published by the leader.

Under a given pricing scheme, the provider decides on the providing content to maximize its expected revenue. For the follower i, the utility function is composed of the benefits and costs of providing content to the content consumer.

$$U_i = \frac{f_i}{\sum_{j \in \text{Followers}} f_j} \cdot p - f_i c_i \tag{4.19}$$

where f_i denotes the total providing content, and c_i denotes the cost of unit content.

The goal of game theory is to find solutions to game problems so that participants eventually reach a stable equilibrium. Nash equilibrium is a solution to the non-cooperative game. We first prove that Nash equilibrium exists in the Stackelberg game model proposed in this chapter.

Lemma 1 *There is a Nash equilibrium in the Stackelberg game model proposed in this chapter.*

Proof In the Stackelberg model, the follower i determines the data content scheme according to the total price p published by the leader. The follower's content scheme is a bounded closed set in Euclidean space, and the utility function is continuous in its scheme space, and the first order partial derivative is

$$\frac{\partial U_i}{\partial f_i} = -\frac{f_i p}{\left(\sum_j f_j \right)^2} + \frac{p}{\sum_j f_j} - c_i \tag{4.20}$$

The second order partial derivative is

$$\frac{\partial^2 U_i}{\partial f_i^2} = -\frac{2p\left[\sum\limits_j f_j - f_i\right]}{\left(\sum\limits_j f_j\right)^3} < 0 \tag{4.21}$$

Therefore, the follower's utility function is a strictly concave function, which ensures that a Nash equilibrium point exists in the Stackelberg game model proposed in this chapter.

We exploited distributed iterative algorithm to solve the optimal pricing for content offloading problem. We predict users' total content consume based on the average content during the previous days. For a single user, the predicted total content is denoted as Q and the unit cost is denoted as c_i, then the initial total price announced by the leader is $p_0 = Q \cdot \min(c_i)$. By predicting the data content consume of user, we can effectively reduce the number of iterations in the game model, thus improving the efficiency of the algorithm.

In the two-stage Stackelberg game, at each moment τ, the leader announces the pricing scheme p to followers. According to the total price p, followers seek to find the optimal content scheme by solving the multiple quadratic equation.

$$\begin{cases} -\dfrac{f_1 p}{\left(\sum\limits_j f_j\right)^2} + \dfrac{p}{\sum\limits_j f_j} - c_1 = 0 \\[4mm] -\dfrac{f_2 p}{\left(\sum\limits_j f_j\right)^2} + \dfrac{p}{\sum\limits_j f_j} - c_2 = 0 \\[2mm] \qquad\qquad \vdots \\[2mm] -\dfrac{f_n p}{\left(\sum\limits_j f_j\right)^2} + \dfrac{p}{\sum\limits_j f_j} - c_n = 0 \end{cases} \tag{4.22}$$

According to the multiple quadratic equation, the optimal content scheme is

$$f_i = \frac{(n-1)p}{\sum\limits_j c_j}\left(1 - \frac{(n-1)c_i}{\sum\limits_j c_j}\right) \tag{4.23}$$

If the utility of the leader also reaches the maximum at this time, all participants in the game have reached the Nash equilibrium. Otherwise, at the next moment, the leader continues to iterate on the pricing scheme according to Eq. (4.8), where λ is the adjustment step size, and announces the new price p to followers. The iterating stops until the utility of leader reaches its maximum.

$$p(\tau + 1) = p(\tau) + \lambda \tag{4.24}$$

The pseudocode of the iterative algorithm for solving Nash equilibrium is shown in Algorithm 1.

Algorithm 1 Distributed content offloading and pricing

Input: records: recorded information
Output: p: pricing scheme, f: content scheme
1: Predict user opportunistic content provider based on Random Forest;
2: For each user u_i in user set U, calculate mobile user collection that may access the same base station at the same time period as user u_i: $E(u_i) = \{v_1, v_2, \ldots v_{m(u_i)}\}$;
3: For each user v_j in $E(u_i)$, filter on the content dimension by determine whether the interest set of user u_i and user v_j is an empty set, obtain the content providers collection $\phi(u_i)$ of user u_i;
4: Calculate the predicted total content Q of user u_i after content consume analysis.
5: Calculate the initial total pricing $p_0 = Q * \min(c_i)$ of user u_i;
6: Calculate the utility function $U_0(\tau)$ of user u_i, while $U_0(\tau) <= \max\{U_0\}$, jump to Step-7 and Step-8, otherwise end while.
7: Iterate over the total pricing p according to Eq. (4.8), for each follower in $\phi(u_i)$ calculate the content scheme f_i according to Eq. (4.7);
8: $\tau = \tau + 1$, jump to Step-6.

In the previous section, the user's opportunity content provide in the coming day can be predicted, thereby obtaining content that the BS and the content providers can provide to the content consumer. Then, the mobile users' information is utilized for dynamic content offloading decision.

However, in realistic network situations, the system state information may not be effectively obtained in dynamic network situations. The information of the content provide and content consumer may be unpredictable or difficult to predict accurately. In this case, the user content consume information for the next day is unknown, only the user history opportunity content provide and content consume records cloud be utilized.

In this section, we consider how to formulate an online content offloading scheme in dynamic network situations, so as to maximize the revenue of multiple participants such as BS, content providers, and content consumers. Based on the reinforcement learning, we use the gradient bandit algorithm to analyze the historical opportunistic content provide and make optimal content offloading decisions through multi-day long-term learning training.

Since the content offloading scheme has a close relationship with the content consume in each time period, the user content consume, which is the total content load of the BS in dynamic network situations, needs to be analyzed first. After analyzing the total content consume, we analyze the user history opportunity content provide and obtain the content that can be offloaded in the dynamic network.

The content provider caches the content data that has been downloaded and browsed on the terminal device, and waits until the mobile user in the vicinity requests the content, and directly transmits the content data to the content consumer by using a short-distance communication method such as WiFi or Bluetooth. Users do not need to pass backhaul, which can save the bandwidth of the backhaul link. To perform the opportunistic content provide, the following three conditions must be met: the users need to have an intersection in time, location, and content dimensions. First, only when the data streams of the two users have an intersection in the communication time, can they perform opportunistic content provide; second, the user pairs have an intersection in the location dimension, since WiFi and Bluetooth can only support communication within a short distance, so only when two users are in the proximity range, the content can be offloaded through the opportunistic content provide; then, the users have common interests in the content and browse the same type of network content data. Content provider caches the content data that it has browsed on the mobile phone. When the content consumer requests this type of content data, the opportunistic content provide can be performed between user provider and user consumer. According to the historical interest content, the content that can be offloaded in the dynamic network is analyzed.

In the user dynamic moving network, a user pair set having an intersection of time and location dimensions in each time slot in each day is counted at first, and the user set in the t time period is as follows:

$$\pi\left(t\right) = \left\{\phi_1, \phi_2, \ldots, \phi_{n(t)}\right\} \tag{4.25}$$

where $\phi_k = \left(u_i, u_j\right)$ denotes the user pair who has an intersection of time and location, and $n\left(t\right)$ denotes the number of user pairs in the t time period. For each user pair, if there is an intersection of the interest content the two users, it is considered that the content of the user u_j consumes is cached in the mobile phone of user u_i, and the content offloading can be performed. Finally, the content consumption of all available content providers of the t time period is summed to obtain the offloaded content during this period.

After we design utility functions for the BS, the content provider, and the content consumers in the realistic network situations, we consider using the gradient bandit algorithm in reinforcement learning to solve the online content offloading scheme that optimizes the system's total utility objective function.

The goal of reinforcement learning is to learn how to act based on the environment to maximize the expected benefits. The gradient bandit algorithm was exploited to solve the optimal mobile data online content offloading scheme. In the model, there are two kinds of actions. When the action is 1, it means that the content is provided by the BS, with the corresponding benefit $R_t\left(1\right) = U\left(x_t = 0\right)$. When the action is 2, the part of content is obtained from the content provider, with the corresponding benefit $R_t\left(2\right) = U\left(x_t = 1\right)$. The reference factors for these two kinds of actions are denoted as $H_t\left(1\right)$ and $H_t\left(2\right)$. At the beginning $t = 0$, the probability of selecting the two types of actions is

$$\Pr\left(A_t = 1\right) = \pi_t\left(1\right) = \frac{e^{H_t(1)}}{e^{H_t(1)} + e^{H_t(2)}} = \frac{1}{2} \tag{4.26}$$

$$\Pr\left(A_t = 2\right) = \pi_t\left(2\right) = \frac{e^{H_t(2)}}{e^{H_t(1)} + e^{H_t(2)}} = \frac{1}{2} \tag{4.27}$$

In subsequent iterations of t rounds, π_t (1) and π_t (2) are updated following the steps.

The preference factor H_t (a) is gradient iterated according to the following formula:

$$H_{t+1}(a) = H_t(a) + \alpha \frac{\partial E[R_t]}{\partial H_t(a)} \tag{4.28}$$

where $E[R_t]$ denotes the expected return value, $E[R_t] = \sum_b \pi_t(b) q_*(b)$, and $q_*(b) = E(R_t/A_t = b]$ denotes the expected benefit when the behavior action is b.

Further, the partial derivative is expanded into the following form, because when the preference factor H_t (a) changes, the probability of some behaviors rises, and the probability of other behaviors decreases, but the sum of the probabilities remains 1, so the sum of the partial derivatives on all behaviors is 0, $\sum_b \frac{\partial \pi_t(b)}{\partial H_t(a)} = 0$. So a scalar X_t that doesn't depend on b is introduced.

$$\frac{\partial E[R_t]}{\partial H_t(a)} = \frac{\partial}{\partial H_t(a)} \left[\sum_b \pi_t(b) q_*(b) \right]$$

$$= \sum_b q_*(b) \frac{\partial \pi_t(b)}{\partial H_t(a)} \tag{4.29}$$

$$= \sum_b (q_*(b) - X_t) \frac{\partial \pi_t(b)}{\partial H_t(a)}$$

The above formula is then rewritten as a mathematically expected form, in which $X_t = \overline{R}_t$, because $E[R_t/A_t] = q_*(A_t)$ and R_t is irrelevant to other items, $q_*(A_t)$ can be replaced by R_t.

$$\frac{\partial E[R_t]}{\partial H_t(a)} = \sum_b \pi_t(b)(q_*(b) - X_t) \frac{\partial \pi_t(b)}{\partial H_t(a)} / \pi_t(b)$$

$$= E \left[(q_*(A_t) - X_t) \frac{\partial \pi_t(A_t)}{\partial H_t(a)} / \pi_t(A_t) \right] \tag{4.30}$$

$$= E \left[(R_t - \overline{R}_t) \frac{\partial \pi_t(A_t)}{\partial H_t(a)} / \pi_t(A_t) \right]$$

According to the quotient's derivative formula, $\frac{\partial \pi_t(b)}{\partial H_t(\alpha)}$ is expanded into the following form, in which $\delta_{a=b}$ is defined as when $a = b$, $\delta_{a=b} = 1$, otherwise $\delta_{a=b} = 0$.

$$\frac{\partial \pi_t (b)}{\partial H_t (\alpha)} = \frac{\partial}{\partial H_t (a)} \pi_t (b)$$

$$= \frac{\partial}{\partial H_t (a)} \left[\frac{e^{H_t (b)}}{\sum_{c=1}^k e^{H_t (c)}} \right]$$

$$= \frac{\frac{\partial e^{H_t (b)}}{\partial H_t (a)} \sum_{c=1}^k e^{H_t (c)} - e^{H_t (b)} \frac{\partial \sum_{c=1}^k e^{H_t (c)}}{\partial H_t (a)}}{\left(\sum_{c=1}^k e^{H_t (c)} \right)^2} \tag{4.31}$$

$$= \frac{\delta_{a=b} e^{H_t (b)} \sum_{c=1}^k e^{H_t (c)} - e^{H_t (b)} e^{H_t (a)}}{\left(\sum_{c=1}^k e^{H_t (c)} \right)^2}$$

$$= \pi_t (b) (\delta_{a=b} - \pi_t (a))$$

According to the derivation of the above formula, the following formula can be obtained.

$$\frac{\partial E [R_t]}{\partial H_t (\alpha)} = E \left[\left(R_t - \overline{R}_t \right) \pi_t (A_t) \left(\delta_{a=A_t} - \pi (a) \right) / \pi_t (A_t) \right] \tag{4.32}$$

$$= E \left[\left(R_t - \overline{R}_t \right) \left(\delta_{a=A_t} - \pi_t (a) \right) \right]$$

Substituting the above formula into Eqs. (4.28) and (4.33) can be obtained.

$$H_{t+1} (\alpha) = H_t (\alpha) + \alpha \left(R_t - \overline{R}_t \right) \left(\delta_{a=A_t} - \pi_t (a) \right) \tag{4.33}$$

Assume that the current action at time t is A_t, the iteration is as Eq. (4.34). The iteration for action $a \neq A_t$ is Eq. (4.35).

$$H_{t+1} (A_t) = H_t (A_t) + \alpha \left(R_t - \overline{R}_t \right) (1 - \pi_t (A_t)) \tag{4.34}$$

$$H_{t+1} (a) = H_t (a) - \alpha \left(R_t - \overline{R}_t \right) \pi_t (a) \tag{4.35}$$

The probability of selecting action a is as follows:

$$\pi_t (a) = \frac{e^{H_t (a)}}{e^{H_t (1)} + e^{H_t (2)}}, \quad a \in \{1, 2\} \tag{4.36}$$

R_t is the reward of selecting action A_t at time t, and $\overline{R}_t$ is the average reward of selecting action A_t in the previous t round. At time $t+1$, we choose to obtain content from the BS or content provider according to the probability $\pi_t (1)$ and $\pi_t (2)$. $\overline{R}_t$ is calculated as follows:

$$\overline{R}_t = \frac{\sum_{i=1}^{t-1} R_i \delta_{A_i = A_t}}{\sum_{i=1}^{t-1} \delta_{A_i = A_t}} \tag{4.37}$$

Using $\overline{R}_t$ as a benchmark for comparing returns, according to Eqs. (4.34) and (4.35), for behavior a, when the return value of selecting behavior A_t at time t is greater than $\overline{R}_t$, the probability of selecting behavior A_t increases, and when the return value of the selecting behavior A_t at time t is less than R_t, the probability of selecting behavior A_t is attenuated. Conversely, for the behavior a, when the return value of selecting behavior A_t is greater than $\overline{R}_t$ at time t, the probability of selecting behavior a decreases, and when the return value of selecting behavior A at time t is less than $\overline{R}_t$, the probability of selecting behavior a increases.

We use the gradient bandit algorithm in reinforcement learning to solve the problem. The pseudo code of the online content offloading algorithm based on gradient bandit is shown in Algorithm 2.

Algorithm 2 Online content offloading algorithm based on gradient bandit

---
Input: q_t: Total content record information, f_t: Offloaded content record information;
Output: $\pi_t(a)$: Probability of choosing behavior a;
1: **for** $i = 1$ to T **do**
2:     Calculate the profit value of the selection behavior
3: **end for**

Our simulation is based on real dataset provided by China Mobile Communication Corporation. The dataset consists of a 24 day mobile Internet detail records of users(UDRs), covering all the Internet records by 1,614,291 users in a 24 day, from November 21st, 2014, to December 13th, 2014.

To verify the method, we select active users with more than 20 records from the original dataset. The size of our dataset is shown in Table 4.1. The description of user data flow is shown in Table 4.2.

In this part, we select the data stream from December 1, 2014, to December 8, 2014, as the training set and the data stream on December 9, 2014, as the test set. We use accuracy and recall as evaluation indicators for the forecasting algorithm.

When the user browses the content on the Internet using their device, the relevant data service information generated by the user's Internet access is stored. The information includes the obfuscated user number, the number of the base

Table 4.1 Size of dataset

Name	Values
Number of records	42,023,038
Number of users	1,614,291
Number of BSs	10,787
Time range of records	2014/11/21 00:00:00 to 2014/12/13 23:59:59

Table 4.2 The description of user data flow

Field	Description
Uid	42,023,038
Number of users	1,614,291
Number of BSs	10,787
Time range of records	2014/11/21 00:00:00 to 2014/12/13 23:59:59

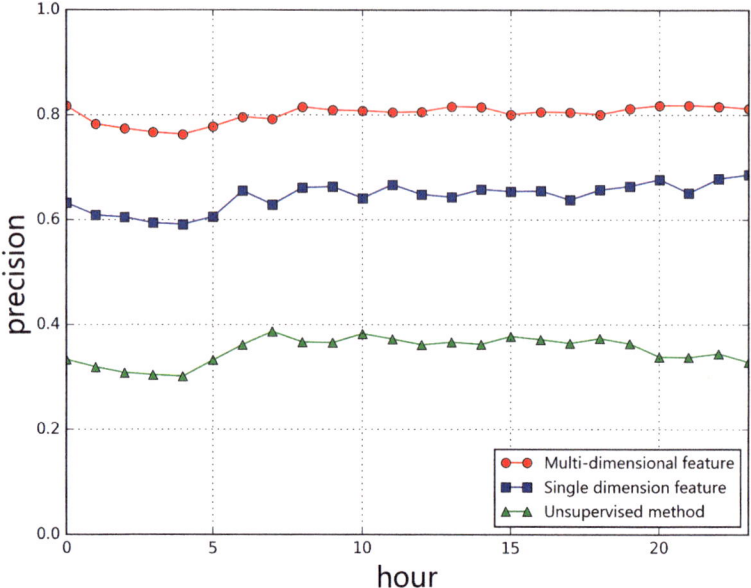

Fig. 4.3 Precision of user content provide prediction algorithm

station in which it is located, the start and end time that user accessed the base station, upstream and downstream content that user consumed, and resource URL currently accessed. Each data stream record can be simplified as an eight tuple:<Uid,Longitude,Latitude,Stime,Etime,Upload,Download,host>, where the descriptions of each field are shown in Table 4.2.

Figures 4.3 and 4.4 plot the accuracy and recall rate for different time periods, respectively. In the figures, the abscissa is in hours, the ordinate is the accuracy and recall rate of the user content providing prediction algorithm in the corresponding time period. Compared with the algorithm that only considering the traditional network structure features and unsupervised algorithm, this algorithm significantly improved the accuracy of the prediction, which remained at around 0.8 at all times.

In the random forest model, the split index of the tree is Gini index. To obtain the importance of each attribute, we average the characteristics' Gini index of multiple decision trees. Figure 4.5 shows the importance of all features, which is sorted in descending order. The value of each feature in the vertical axis is

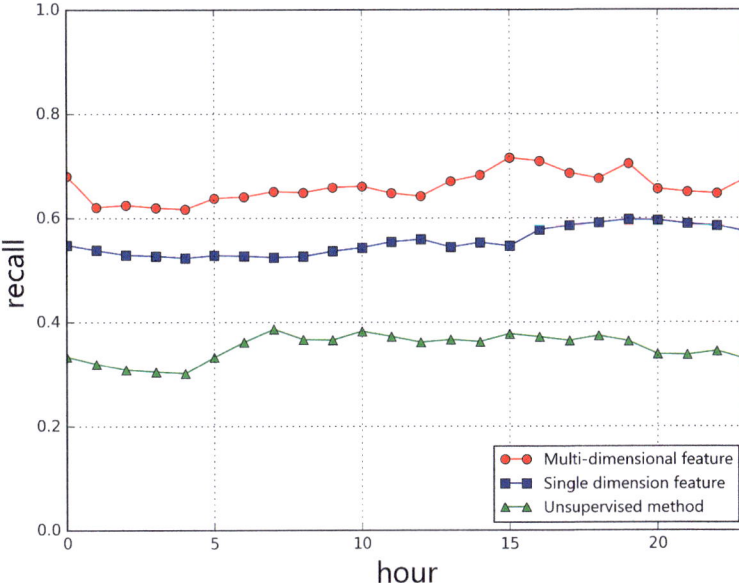

Fig. 4.4   Recall of user content provide prediction algorithm

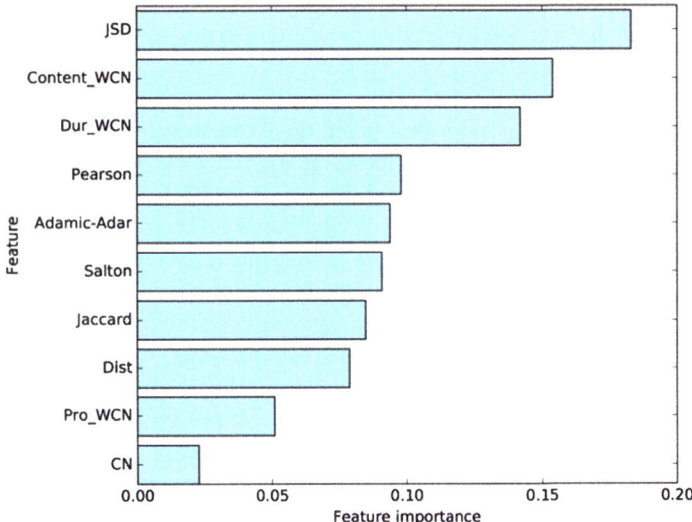

Fig. 4.5   Distribution of feature importance in random forest models

the feature importance score in the user content provide prediction algorithm. As
can be seen from the figure, indicators based on the user's Internet duration and
content attributes, user space similarity have an important influence on whether
users can share interest content with each other. The degree of importance of

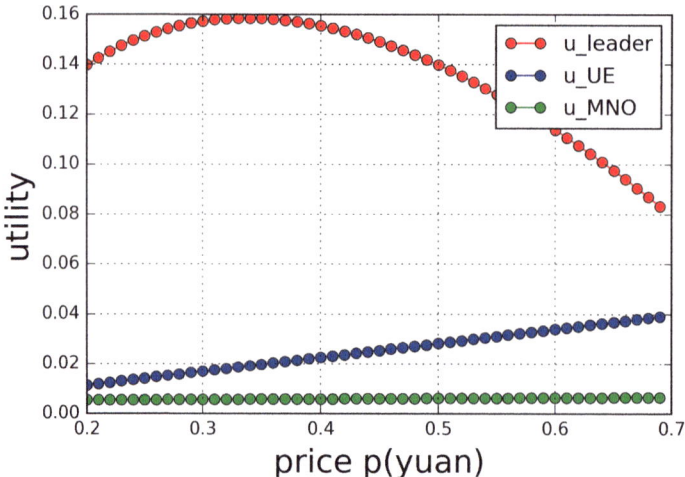

Fig. 4.6 Utility function change

individual behavior features is ranked high, and the feature similarity between nodes is significantly higher than the general features, indicating that these features have a greater impact on the opportunistic content provide between individuals, so taking these features into consideration to a certain extent can help to improve the prediction performance.

The data records from 2014/12/01 to 2014/12/09 are used to verify the effectiveness of data offloading algorithm based on Stackelberg game. Take a content consumer user with five followers including the edge device and four content providers as an example. The predicted data content of him is 1 Mb. As shown in Fig. 4.6, when price p is increasing from the initial price, the utility of the leader gradually increases. When $p^* = 0.34$, all participants reach the Nash equilibrium in the Stackelberg model. Figure 4.7 shows the content provided by content providers changing with different price. With the increasing price, followers are encouraged to provide more data content for the content consumer user.

We further compared the number of iterations of the algorithm proposed in this chapter with the compared method in which the initial price is 0. As shown in Fig. 4.8, compared to the original Stackelberg game pricing, our method sets the initial price according to the prediction, which can quickly converge to the optimal pricing with less number of iterations. The efficiency of the algorithm has been improved efficiently.

Figure 4.9 shows the data content load of the BS after applying the mobile data offloading scheme in the 24-h period. During the peak period of content consumes, the 22nd time period, the amount of offloaded data content accounts for 34.5% of the original content load. The experiment shows that the mobile data offloading algorithm based on Stackelberg game can effectively reduce the data content pressure of the BS.

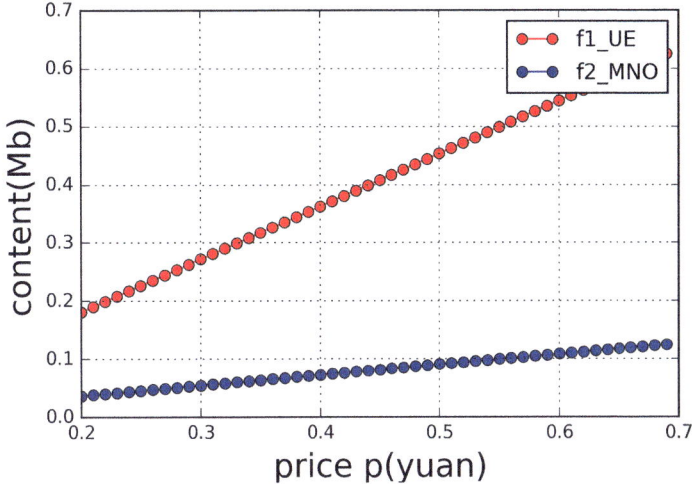

Fig. 4.7 Data content scheme change

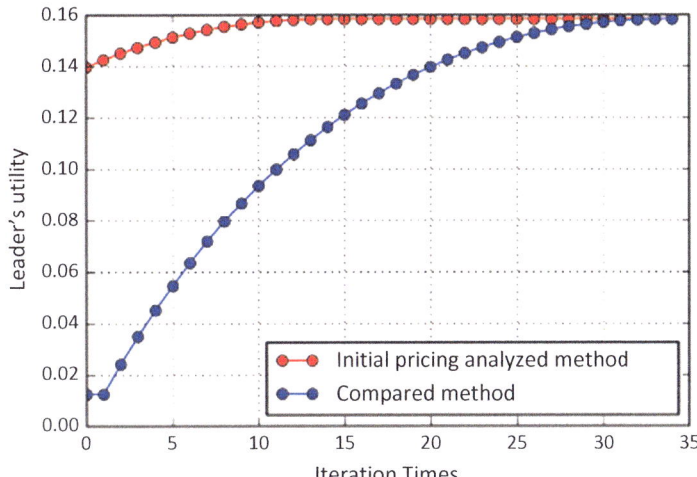

Fig. 4.8 Iterations numbers

We applied the online content offloading algorithm based on reinforcement learning to the real cellular network data. The part of content of users who have content provide behavior and the intersection of history interest content is the possible offloaded content. Figure 4.10 shows the total content consumes and possible offloaded content in 22 days, respectively. We can observe the tidal phenomenon of data content during the period.

Figure 4.11 shows the offloading percentages with different weight parameters. In the figure, the abscissa represents the weight of the content provider's utility

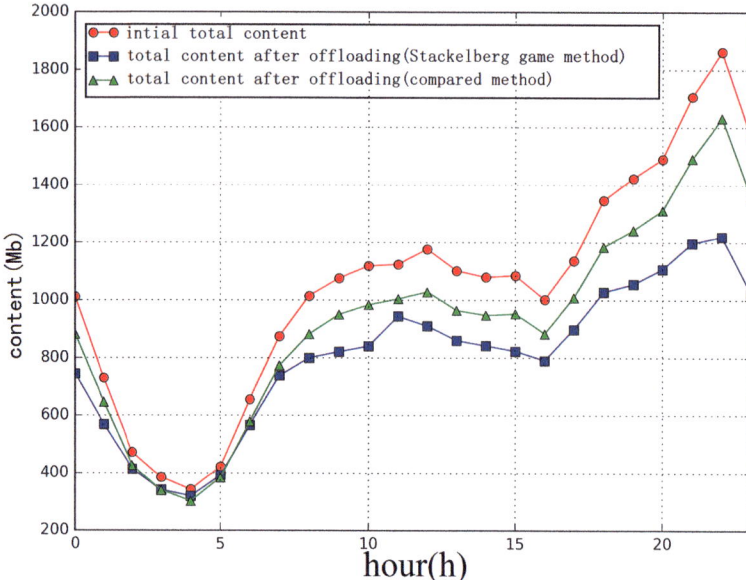

Fig. 4.9 Content load of BS

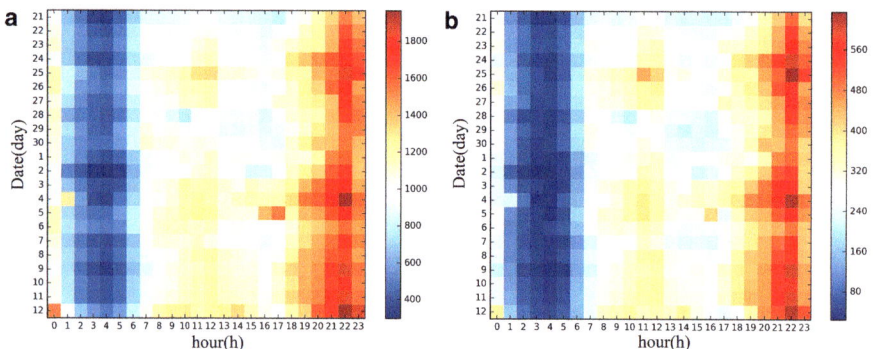

Fig. 4.10 Data content in 22 days. (**a**) shows total data content consumes. (**b**) shows possible offloaded data content

function, the left ordinate represents the weight of the BS's utility function, and the right ordinate represents the proportion of the offloaded content. As the weighting coefficient given to the BS's utility function increases, the probability of selecting the execution of the offloading has a decreasing trend, the proportion of the offloaded content decreases.

Considering the utility of all parties, we take $\lambda_1 = 0.3$, $\lambda_2 = 0.3$, and $\lambda_3 = 0.4$ to make the weights of the three parties relatively balanced. When the weight parameter is set as $\lambda_1 = 0.3$, $\lambda_2 = 0.3$, and $\lambda_3 = 0.4$, the content load after applying content offloading scheme is as shown in Fig. 4.12. During 8–16 h and 18–23 h,

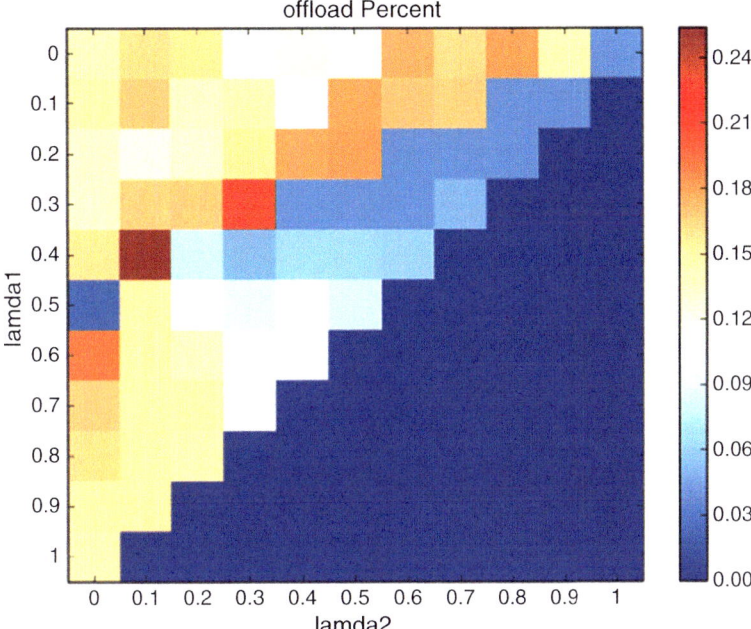

Fig. 4.11 Offloading percentages with different weight parameters

which are the peak period of content consumes, users choose to perform content offloading. During 2–7 h, which is the trough period of content, users still choose to obtain content from the BS. The offloaded content accounts for 22.9% of the original content of the BS. The experiment shows that our online content offloading algorithm based on reinforcement learning can effectively reduce the content load of the BS.

For the future D2D paradigm, the large-volume data content of repeated content request in the networks will burden the backhaul. Considering the content visit shows great pattern with popularity and repetition, we propose two content offloading schemes to deal with network dynamics incurred by mobile users and imperfect system information, respectively. In the first scheme, we propose the dynamic content offloading decision formulated by Stackelberg game among different content providers of D2D server and mobile users. In the second scheme, we propose the online content offloading scheme by using gradient bandit. The evaluations are based on the real dataset from China Mobile Communication Corporation. Both of the schemes provide good efficiency that can alleviate the network backhaul. The dynamic content offloading decision scheme provides flexible and low-cost offloading with fast convergence. Our online content offloading scheme is with good efficiency in terms of content load.

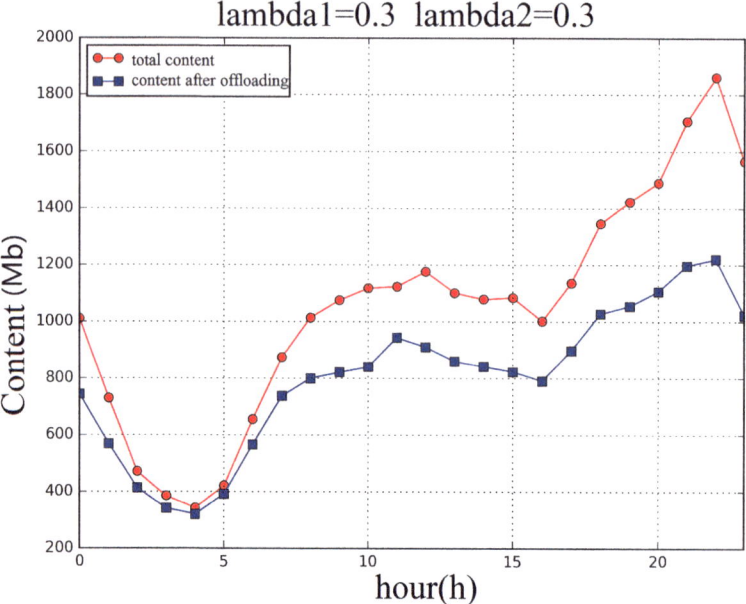

Fig. 4.12 Content load after applying online content offloading scheme based on reinforcement learning

4.2   Mobile Data Application in Green Communication Networks

4.2.1   Green Communication Networks: Concepts and Application

With the advent of the 5G era, the demand for service applications and user experience has forced network operators to find new technologies to reduce operating cost and increase data transmission rate, thereby improving the quality of user experience. The exponential growth of data services has dramatically increased the demand for network facilities and has also caused large energy consumption. And the 5G communication network has been moving toward the deployment of very large-scale and dense base stations (BSs) [31]. Dense BS deployment results in excessive energy consumptions and carbon emissions, which causes hard burden on electrical energy facilities and affects the ecosystem greatly. Green communication, as one of the important requirements of 5G, requires that the total energy consumption of the network will not increase when the user data traffic increases by a thousand times. It is expected that by 2020, the end-to-end energy consumption per bit in the future 5G network will need to be reduced to 1/1000. The IMT-2020 proposes that the future 5G will also work to improve the operational energy consumption and cost efficiency of network construction [32]. Therefore, in the future, 5G communication will be necessary to reduce energy consumption while maintaining various types of business growth. The increase in energy consumption costs has also led people to pay more and more attention on ways to reduce energy consumption.

Due to people's living habits and mobility characteristics, the traffic patterns of BSs in cities show tidal phenomena and spatial differences, that is, traffic loads of base stations experience significant peak and trough periods, and there may also be significant differences in the traffic loads of different base stations at the same time. According to the dense deployment of cellular base stations in cities, it is energy efficient that can effectively improve BS utilization by appropriately turning off some BSs and offload users to nearby BSs during non-peak times.

The important difference compared to the 4G research idea is that 5G is considered to be the first important user experience as a research core. In other words, in the upcoming 5G era, the most important thing is not speed, but more application and user experience, and a more adaptive business model. However, experience of users who are offloaded to other base stations may be affected to some extent. Traditional BS energy-saving strategy ensures the quality of service (QoS) of users as a prerequisite, and mainly concludes service-related indicators such as delay, throughput, jitter, and packet loss while ignoring the influence of business characteristics on user experience. Quality of experience (QoE) is a measure of acceptability based on user acceptance, Zhao et al. [33] pointed out that QoE can reflect user's satisfaction degree toward the network services based on quantitative modeling of QoS. In the multi-service oriented application architecture, accurate assessment of user experience toward different services according to service characteristics and demands for transmission resources are helpful to the optimization of system resources. To solve the problem, Gomez et al. [34] provided a detailed introduction to the QoE evaluation methods of services such as voice, video streaming, and web browsing, and Khan et al. [35] used the utility function from economics to evaluate the specific business services.

QoE overcomes the deficiencies of QoS metrics that ignore service features, it can better describe users' satisfaction with applications or services. Furthermore, considering that the time complexity of solving the problem is growing exponentially with the number of base stations, in this chapter we propose a distributed multiple-service user-experience energy-saving algorithm for base station (DMUES) adopting QoE utility function to evaluate the multi-service user experience, and nonlinear integer programming is utilized to model the energy-saving strategy. And the community is divided into base stations in the city, and the base station energy-saving strategy is implemented separately in each community. Through experiments, it is proved that the proposed DMUES achieves good user experience and energy saving. Furthermore, it greatly reduces the computational complexity and time complexity.

In the urgent need to achieve energy saving, the green cellular network has become a hot topic for researchers. The traditional energy-saving measures of cellular networks are mainly divided into four categories, namely improving the energy efficiency of hardware components, optimizing the energy efficiency of the wireless transmission process, base station hibernation/cell zooming, and deploying heterogeneous units [36]. The way to improve the energy efficiency of hardware components requires network operators to replace cellular network system components on a large scale with high implementation costs. Improving energy efficiency by optimizing the wireless transmission process requires a compromise between energy saving and network performance. The method of deploying heterogeneous units is to reduce energy consumption through the use of plug-and-play small cell base stations in the middle of cellular networks (including microcells, picocells, etc.) [37].

Cell zooming is a cell-like breathing concept that changes the base station's coverage area by adjusting the state of the BS. It includes measures such as adjusting base station height, power, sleep, and full shutdown. When a base station is in a low-load state, it can serve users under neighboring base stations by increasing its coverage area and reducing the load on neighboring base stations. Even if all the base stations are under low load, even partial base stations can be completely turned off to achieve energy saving. The BS consumes the largest proportion of energy in the cellular network. By monitoring the traffic of the base station, selectively shutting down some base station transmitting units and cooling equipment (air conditioners, etc.) during off-peak hours, the energy-saving effect is very considerable. In 5G heterogeneous networks, the deployment of dense cellular network base stations makes the coverage area of a single base station smaller. The base station's business model is more random, which makes the strategy of shutting down the base station more desirable. Combining 5G heterogeneous network deployment with the base station's sleep mode will yield significant benefits in terms of energy savings [38].

The base station closure policy usually determines the state of the communication unit (base station) of the cellular network by monitoring the traffic load in the network. It is mainly divided into three categories, which are strategies based on traffic load, strategies based on user association, and strategies combining traffic load and user association, respectively. Han et al. [39] proposed base switching strategy based on traffic load, and solved the problem by centralized greedy decision-making and decentralized autonomous decision-making, respectively. Zhu et al. [40] proposed a QoS-aware user association scheme to reconfigure the cell association for energy saving. Oh et al. [41] proposed a practically implementable switching-on/off base energy-saving algorithm considering base load and user association both, and the algorithm can be operated in a distributed manner with low computational complexity. Jiang et al. [42] estimated the aggregate traffic demands of the BS communities and proposed a switch-off strategy while guaranteeing minimal service requirements. Son et al. [43] developed a theoretical (and also practical) framework for BS energy saving that encompasses both dynamic BS operation and user association.

The load based BS on–off switching strategy pays more attention to energy efficiency other than the association status of users. The user association switching strategy needs to switch the BS on–off state frequently according to user association, which brings extra energy consumptions. The hybrid method of joint load and user association based switching strategy avoids frequent status switching with more balanced energy consumptions and better user experience.

For multi-service application architectures, different services have different characteristics and requirements for network transmission quality. The models of users' perceptions of different services are different and need to be evaluated using the corresponding models. This will help optimize the resources of the system. The cellular network is a multi-service application architecture. For the characteristics of the business and the demand for network transmission resources, accurate assessment and modeling of different services can help to optimize system resources. The European Telecommunications Standards Institute introduced the concept of QoE, and provided a mapping analysis between QoE and network performance indicators for different services [44]. The literature [45] proposes a general formula that links QoE and QoS parameters through an exponential relationship.

4.2.2 Multiple-Service User-Experience Energy Saving for Green Communication Networks

In order to describe the energy consumption of the BSs, this section gives the energy consumption model. A BS usually consists of several cells with different radiation azimuths, and cells share the base architecture like cooling system of the BS. Energy consumption of a cell is divided into static consumption (consumption from transmission antenna and power amplifier) and dynamic consumption (consumption related with load/utilization, which varies with cell load), and cells of a BS share the consumption of base architecture.

For the measurement of user experience for specific service types in mobile cellular networks, this chapter introduces QoE utility function to model user experience for multi-services.

To get users association state to cells, we should first get the coverage area of each cell. Each BS has multiple cells, and the cells belonging to the same BS have the same geographical coordinate. The coverage is different. The coordinate of a point closer to the BS on the center line of the covered cell area with the sector as the position coordinate of the cell is the azimuth, and is a very small positive value. Then we adopt Voronoi tessellation to divide the areas based on the location (latitude and longitude) and antenna radiation azimuth of cells. Cell division example from a typical city of Zhejiang province, China, is showed in Fig. 4.13.

Whether a user can access to a cell depends on user's distance and direction relative to the cell, user–cell affinity relationship, and the affinity graph. If a user is within the coverage area of a cell, then we can add an edge between them.

In this chapter, user association strategy proposed in [46] is adopted. It is assumed that the inter-cell interference and environmental interference are both Gaussian

Fig. 4.13 Coverage areas
division of cells

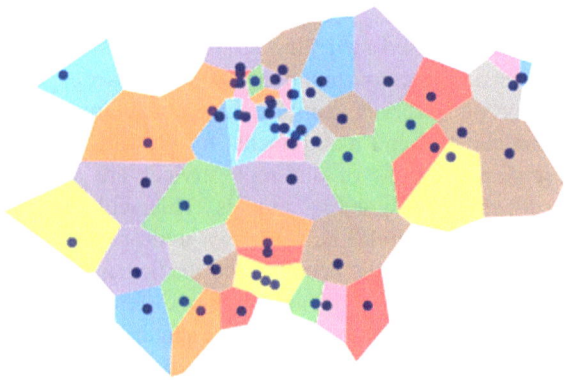

white noises, and the received signal strength of user is only related to the distance. On this condition, users will associate to the nearest active cell it can be associated with. For example, users under cell 1 can be associated with cell set, when cell 1 is closed, users under it will be associated with the active cell 7 from which they can receive the best average signal strength.

In a BS switching strategy, users under a closed sector are associated with neighboring cells. In a BS switching strategy, users under a closed cell are associated with neighboring cells. A cooperative relationship between cells is represented by a directed weighted graph. V is the set of sector nodes, and E is the set of edges of nodes. When 80% of users under a cell are under the coverage area of a cell, there are edges in the graph: the size of the edge is the average distance between the user and the sector under the cell. After the cell is closed, users under the sector can associate with the set of cells.

Each BS in the cellular network is consisted of multiple cells with different radiation azimuths, cells share the base stations infrastructure such as the refrigeration equipment. Energy consumption of a cell is consisted of static energy consumption and dynamic energy consumption. Static energy consumption concludes consumption of power amplifiers, antennas, etc. Dynamic energy consumption is needed for processing of the traffic loads, and is positively related to load ρ. Assuming that $\{\rho_1, \rho_2, \rho_3, \ldots, \rho_n\}$ represents the traffic load demands of cells, and $\{\rho_1^*, \rho_2^*, \rho_3^*, \ldots, \rho_n^*\}$ represents cells current traffic loads with the energy-saving strategy. Power consumption of cells increases linearly with the traffic load:

$$P_{\text{cell}}(\rho) = k \cdot \rho + c \qquad (4.38)$$

k is the coefficient, and c is the static energy consumption of the cell.

Consumption of a base station is

$$P_{\text{base}} = \sum_{\text{cell} \in \text{base}} P_{\text{cell}}(\rho_{\text{cell}}) + C \qquad (4.39)$$

Among the function, C is the consumption of the base architecture of the BS.

QoE was proposed by the ITU-T for measuring user's subjective acceptability to applications or services perceived. For the mapping from objective measurements (in terms of QoS) into subjective metrics (in terms of QoE perceived by the user), we adopt the utility function toward multi-services which is defined as the gain users obtained from a service according to the resource requirements. Mobile cellular network services are mainly classified into four types: VoIP services, streaming services, interactive services, and background services. Each type of services has specific characteristic and different transmission resource requirement. Therefore, different forms of utility function are required for modeling multi-service user experience.

VOIP: such as audio calls and video calls, is a hard real-time service that requires strict end-to-end guarantee and is particularly delay sensitive. Its utility function is a unit step function of the allocated resource r. When the allocated resource r is smaller than the minimum transmission resource requirement r_{min}, the utility value is 0, and when the allocated resource equals to r_{min}, the utility value steps to its max value 1 and will not increase anymore, the utility function is as follows:

$$U_{VOIP}(r) = \begin{cases} 0, r < r_{min} \\ 1, r \geq r_{min} \end{cases} \tag{4.40}$$

Streaming (STM): audio and video streaming services that can use data buffering technology have a certain degree of tolerance for media distortions and can be accepted when packets exceed the latency limitation slightly or even get some loss. The utility function is in exponential form. When the resource r is smaller than a certain smaller value r_{min}, the utility value increases from zero to r, and its growth rate increases. When r gets greater than r_{min}, the growth rate of utility function decreases. Due to the limited system transmission resource r_{max}, the value of the utility function reaches its maximum value when the resource reaches r_{max} and will no longer increase, the utility function is as follows:

$$U_{STM}(r) = 1 - e^{-\frac{k_1 r^2}{k_2 + r}}, r \leq r_{max} \tag{4.41}$$

Interactive Services (WEB): Mainly are web browsing, chatting, games, and other services that have a certain tolerance for delay. The utility function is an exponential function with the allocated resources r. When r is lower than the minimum resource requirement r_{min}, the utility value is 0; when r_{min} gets greater than r_{min}, the utility value steps to a larger value and begins to increase monotonously with r, and the growth rate keeps decreasing. The utility value reaches its maximum when r reaches the limited system transmission resource r_{max}, the utility function is as follows:

$$U_{WEB} = \begin{cases} 0, r < r_{min} \\ 1 - e^{-\frac{k_3 r}{r_{max}}}, r_{min} \leq r \leq r_{max} \end{cases} \tag{4.42}$$

Background BK: Services such as file transferring and email, which has high tolerance of delay. The utility function is an exponential function of resource r and increases with the resource r with decreasing growth rate. When the allocated resource r reaches the system's maximum resource limit r_{max}, the utility function will reach its maximum value

$$U_{BK}(r) = 1 - e^{-\frac{k_4 r}{r_{max}}}, r \le r_{max} \tag{4.43}$$

Among them, k_1, k_2, k_3, k_4 are positive model parameters used to adjust the growth rate of utility function. The range of value for utility function is [0,1].

The average transmission rate received by users under cell i is: $r_i = W \log(1 + \text{SINR})$, where W is the channel bandwidth, and $\text{SINR} = \frac{p_i + g_i}{\sum\limits_{m \in N_i, m \ne j} g_m + \sigma_0}$ is the signal to interference plus noise ratio, in which p_i is the transmission power of antenna, σ_0 is white noise from the environment, $\sum\limits_{m \in N_i, m \ne j} g_m$ is the inter-cell interference from the neighbor cells which can also be considered as Gaussian white noise, $g_i = L(d_i) = \left(1 + \left(\frac{d_i}{40}\right)^{3.5}\right)^{-1}$ is the channel gain caused by channel fading and Rayleigh fading, and d_i is the average distance from users under cell i to the cell.

In the last section we have introduced the system model in detail and included the base station energy model and multi-service user experience model. The base station switching strategy is closing some cells and offloading users under them to their neighbor cells according to cells cooperation relationship, and ensuring that the user experience be guaranteed. Our goal is to save as much energy as possible at the expense of a smaller loss of user experience, that is, base station energy saving is a multi-objective optimization problem of base station energy and multi-service user experience.

Assume that the status vector of cells is $\{x_1, x_2, x_3, \dots, x_n\}$, $x_i = \{0, 1\}$, let $\{\rho_1, \rho_2, \rho_3 \dots, \rho_n\}$ be the cell load demand vector and $\{\rho_1{}^*, \rho_2{}^*, \rho_3{}^* \dots, \rho_n{}^*\}$ be the current cell load vector. $\rho_i{}^*$ is related with the status of cell i and status of its neighbors.

From the user-cell affinity graph we can get the cell set N_i that users under cell i can be associated with, and the cell set C_i whose serving users can be associated with cell i. When cell j in C_i is closed, and neighbor cells closer to cell j than cell i are all closed, cell i will be the nearest active cell of cell j, then the load of cell j will be offloaded to cell i, that is, $\rho_i{}^*$ is related to state of N_i and state of neighbor cells of cells in N_i, $\rho_i{}^*$ can be expressed as follows:

$$\rho_i^* = x_i \cdot \rho_i + (1 - x_j) \cdot \sum_{j \in C_i} \rho_j \cdot \prod_{m \in N_j, d_{jm} < d_{ji}} (1 - x_m) \tag{4.44}$$

As cell consumption is related with load, cell consumption will be

$$E_i = P_{\text{cell}}(\rho_i^*) \tag{4.45}$$

And BS infrastructure consumption is related with status of all cells under it

$$E_b = C \cdot \left(1 - \prod_{i \in b}(1 - x_i) \right) \tag{4.46}$$

We adopt the average distance from users random location in the cells coverage area to the cell as the distance from user to its accessing cell. Distance from user u under cell i to its accessing cell d_u is related to the state of cell i and N_i

$$d_u = x_i \cdot d_{ii} + (1 - x_i) \cdot \sum_{j \in N_i} x_j \cdot \prod_{m \in N_i, d_{im} < d_{ij}} (1 - x_m) \tag{4.47}$$

In the equation, d_{ii} is the distance from user u to cell i, and d_{ij} is the distance from user u to its neighbor cell j.

And QoE of user u under cell i whose accessing content type is *cont* is

$$q_u = U_{\text{cont}}(r(d_u)) \tag{4.48}$$

As value of QoE utility function is normalized with a maximum of 1, we define the QoE loss as

$$c_u = 1 - U_{\text{cont}}(r(d_u)) \tag{4.49}$$

User set with different accessing content type under cell i is U_i, sum of QoE loss under cell i is

$$Q_i = \sum_{u \in U_i} c_u \tag{4.50}$$

The purpose of this chapter is to achieve energy savings of BSs with less loss of user experience, so the goal can be transferred into getting the minimum value of the optimization function $\min(f)$ which combines the energy consumptions and QoE losses.

$$\min(f) = \left(\sum_{i \in B} E_i + \sum_{b \in B} E_b \right) / P_i + \eta \cdot \sum_{i \in B} Q_i \tag{4.51}$$

$$\text{s.t.} \rho_i^* < \rho_{\max}, \forall i \in \{1, 2, \dots, n\} \tag{4.52}$$

$$\text{s.t.} x_i + \sum_{m \in N_i} x_m >= 1, \forall i \in \{1, 2, \dots, n\} \tag{4.53}$$

Among them, η is a parameter for adjustment, by adjusting the value of η we can change the proportion of energy consumption and user experience in the optimization. Increasing the value of η means increasing the importance of user experience. Decreasing the value of a corresponds to increase in importance of energy consumption. For example, if η equals to value 1, it indicates that user's optimal experience is equivalent to a cells energy consumption. Equation (4.52) gives the burden limit of cells, and means that traffic load of each cell should not overpass its service capability $\rho_{\max}$. Equation (4.53) gives the user association protection to ensure that each user should be able to connect to at least one cell.

The multi-objective programming problems of energy consumption and user experience are converted into a joint optimization goal min(f) in this chapter. The energy-saving scheme is modeled as a 0-1 integer nonlinear programming problem of the cell status vector $\{x_1, x_2, x_3, \ldots, x_n\}$, $x_i = \{0, 1\}$. And the goal is to get a set $\{x_1, x_2, x_3, \ldots, x_n\}$ to achieve the joint optimization of energy consumption and user experience. And the tradeoff between the energy consumption and the user experience will be achieved.

This chapter uses the classical branch-and-bound algorithm to solve the 0-1 integer nonlinear programming problem. The thought of branch-and-bound algorithm [47] is to relax discrete 0-1 variables into continuous variables range from 0 to 1, and decompose the original problem into disjoint relaxation sub-problems. Target value corresponding to the relaxation solution is taken as the upper bound of the original problem, after iterations, we can get the optimal solution of the original problem. In the end, we use the optimization software Lingo to solve the above 0-1 integer nonlinear programming problem.

The optimization problem in this chapter is a 0-1 polynomial programming problem. Branch-and-bound method solves the deterministic solution of nonlinear integer programming and is more efficient in solving hybrid nonlinear integer programming. The basic idea is to generate a branch and delimitation tree and a sequence of upper and lower bounds. By decomposing the original problem into a series of disjoint relaxed sub-problems, the target value corresponding to the solution of the relaxed sub-problem is taken as the upper bound of the original problem, and the iterative solution to the original problem is obtained [47].

The integer programming problem in this chapter is a convex nonlinear 0-1 integer programming problem, which is a pure integer programming problem. Pure Integer Nonlinear Programing (PINLP).We use an improved linear/nonlinear branch-and-bound method for 0-1 integer programming(LP/NLP based branch-and-bound, LP/NLP-BB).

Relaxation strategies include relaxation of the original problem to continuous space (NLP sub-problems) and linearization of nonlinear constraints (ILP sub-problems).

The optimization goal of Eq. (4.51) can be expressed as follows:

$$
\begin{aligned}
Z_{\text{INLP}} &= \min f(x), \\
\text{s.t.} & \; g(x) \leq 0, \\
& \; m(x) \leq 0, \\
& \; x \in \{0, 1\}^n
\end{aligned}
\tag{4.54}
$$

$g(x) = \{g_1(x), g_2(x), \ldots, g_n(x)\}$ is the nonlinear constraint of Eq. (4.52), $m(x) = \{m_1(x), m_2(x), \ldots m_n(x)\}$ is the linear constraint of Eq. (4.53), $x = \{x_1, x_2, \ldots, x_n\}$.

ILP sub-problems: If the original target Eq. (4.51) is nonlinear, its optimal solution may be the interior point in the convex hull. Such a problem should not be solved directly. Therefore, we introduce an auxiliary variable to convert a nonlinear target into a linear target, and the original objective function is treated as a constraint to obtain the equivalent INLP problem.

$$
\begin{aligned}
Z_{\text{INLP}} &= \min \eta, \\
\text{s.t.} & \; f(x) \leq \eta, \\
& \; g(x) \leq 0, \\
& \; m(x) \leq 0, \\
& \; x \in \{0, 1\}^n, \eta \in R
\end{aligned}
\tag{4.55}
$$

Due to the convexity of the objective function and nonlinear constraints, so at the current point $\hat{x}$, the inequalities $f(x) + \nabla f(\hat{x})^T (x - \hat{x}) \leq f(x)$ and $g(x) + \nabla g(\hat{x})^T (x - \hat{x}) \leq g(x)$ are established. So we can convert constraints to linear constraints, then the original goal is relaxed as follows:

$$
\begin{aligned}
Z_{\text{INLP}} &= \min \eta, \\
\text{s.t.} & \; f(x) + \nabla f(\hat{x})^T (x - \hat{x}) \leq 0 \\
& \; g(x) + \nabla g(\hat{x})^T (x - \hat{x}) \leq 0 \\
& \; m(x) \leq 0
\end{aligned}
\tag{4.56}
$$

NLP sub-problems: Another relaxation relaxes the integer variable of the original problem to continuous space. Assume that each component meets the limit $0 \leq l_I \leq x \leq u_I \leq 1$. The original problem will be transformed into a general nonlinear programming sub-problem.

$$
\begin{aligned}
Z_{\text{NLPR}}(l_I, u_I) &= \min f(x), \\
\text{s.t.} & \; g(x) \leq 0 \\
& \; m(x) \leq 0 \\
& \; 0 \leq l_I \leq x \leq u_I \leq 1
\end{aligned}
\tag{4.57}
$$

If (l_I, u_I) is the upper bound of the feasible domain Eq. (4.54), then the corresponding target value of Eq. (4.57) is to provide an effective lower bound for the original problem. The upper bound of (4.54) is provided by a feasible solution to the relaxed sub-problem. In combination with the linear relaxation of constraints and the continuation of integer variables, when the branch-and-bound method is solved, if an integer point is considered as a feasible solution to the original problem, the original problem becomes a series of LP sub-problems.

$$
\begin{aligned}
&Z_{\text{LP/NLP}} = \min \eta, \\
&\text{s.t.} f(x) + \nabla f^T(\hat{x})(x - \hat{x}) \le 0, \\
&g(x) + \nabla g^T(\hat{x})(x - \hat{x}) \le 0, \\
&m(x) \le 0, \\
&x \in \{0, 1\}^n, \eta \in R
\end{aligned}
\tag{4.58}
$$

Relaxation algorithm provides a basis for the delimitation of the branch-and-bound method. The specific LP/NLP-branch-demarcation algorithm steps are as follows:

First, we use a heuristic algorithm to find the initial feasible solution to the optimization objective function, namely the root node, and then relax the integer variable to $l_I \le \hat{x} \le u_I$.

Step 1:   Select node. Starting from the root node, we select a node in the branch-and-bound tree to solve the loose sub-problem. If this problem is not feasible, we delete this node and search for the child nodes of the branch tree. Otherwise we assume the solution $\hat{x}$.

Step 2:   Pruning. If the value of the objective function corresponding to the solution at this node is greater than the current upper bound, the feasible region of this portion does not contain the optimal solution, and the branch is cut off.

Step 3:   Check integer constraints. If the point $\hat{x}$ does not satisfy the integer constraint, we select a variable that does not satisfy the integer constraint and add the left and right branch constraints $\hat{x}_i = 0$ and $\hat{x}_i = 1$. Otherwise, if $\hat{x}$ satisfies the integer constraint condition and the objective function value is less than the current optimal value, the upper bound is updated, and the branch whose objective function value is greater than the current upper bound is trimmed.

Step 4:   Check if the branch delimiter tree is empty. If the branch and delimitation tree is not empty, then it returns to step 1. Otherwise, the algorithm will be terminated and the current optimal solution will be output.

The branch-and-bound method finds the integer optimal solution in the feasible domain of linear relaxation model of integer programming according to certain search rules. The time complexity of its solution will still increase exponentially with the number of base stations. To implement base station energy-saving strategy in large-scale cellular network clusters, the time complexity is very high. We consider the community division of BS clusters based on the cooperative relationship between cells. BS energy-saving strategy is implemented separately in each community BS to achieve distributed computing to reduce time complexity.

The BSs in the city are deployed according to the functional areas of the city and the users' demands, which are distributed unevenly. We use space cooperation network to describe the cooperation between BS cells in urban space. We use space cooperation network G to describe the cooperation between cells in urban space. The graph is usually expressed as $G = (V, E)$. It consists of nodes and edges. V represents a set of nodes and E represents a set of edges. The network consists of the cell node $< V_i, V_j >$ and the associated edges W_{ij} between the sectors. When there is a cooperative relationship between sectors, increase the edge in. The edge weight is a measure of the tightness between cells and is inversely proportional to the distance $W_{ij} = e^{-d_{ij}^2 / (2 \cdot \mathrm{delta}^2)}$.

When there is a cooperative relationship between sectors, the smaller the distance between the sectors, the greater the connection tightness between them; the greater the distance between base stations, the smaller the connection tightness.

In the connection relationship between base station sectors in the main urban area of Jinhua city, the distribution of base stations in cities is uneven. And the closer we get to the city center, the more dense the base station is. The distance between the BS is close. The connection is more and the network structure has certain community structure. In the suburban areas, the deployment of BSs is sparse and the connection between base stations is sparse, presenting a distinctly small community structure.

The Louvain algorithm is used to divide the network into communities, and the base station energy-saving strategy is implemented separately in each community, which can effectively reduce the time complexity and can be used to save energy in large-scale cellular networks.

In this section, we numerically evaluate the proposed MUES by comparing its performance to several reference algorithms based on the real China Mobile UDR dataset in Jinhua District of Zhejiang province. We use the method of community division to divide the BS clusters in the city. And then we implement energy-saving strategies in each community to reduce the time complexity.

We analyze the cellular network of the urban areas in a city of Zhejiang province, China. Fifty-nine dense cells in a square area are selected for experiment, and the coverage areas of cells are shown in Fig. 4.13. The UDRs used in this chapter are based on users' Internet access behavior which concludes fields of session time, cell location information, URL, traffic load, etc. Table 4.3 shows the main fields and record example of UDRs.

The optimization problem that needs to be solved in the BS energy-saving strategy in this chapter is the 0-1 combination planning problem. The time complexity of the solution is exponentially increasing with the number of targets. First, the community energy-saving strategy optimization model with different sectors is modeled. Then branch-and-bound method is used to solve the optimization problem.

Table 4.3 Field of the usage detail records

userID	Session time	location	URL	Traffic load/B
69201446765	2014-11-21 20:18:24	6893_379A	news.baidu.com	107031

Fig. 4.14 Time logarithmic curve

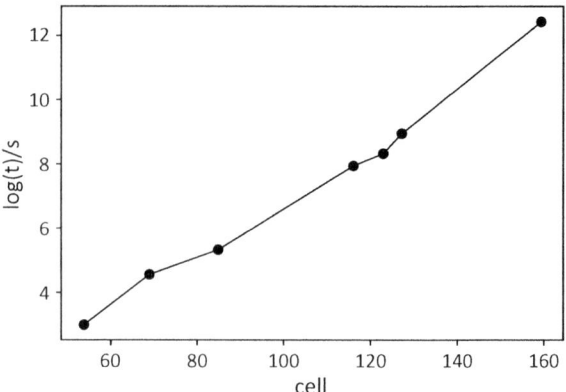

The logarithmic curve of the solution time and the number of cells is shown in Fig. 4.14.

As can be seen in Fig. 4.14, the time complexity of the energy-saving strategy optimization problem is an exponential function of the number of cells: $T(n) = k \cdot 2^n$. The branch-and-bound solver used in this chapter runs on ordinary computer windows systems. When the number of cells is 127, the average optimization solution time reaches 20 min, which is more than 1/3 of the implementation period (1 h) of the energy-saving strategy, and the implementation efficiency is low. When the number of sectors is 160, the complete iterative search process time for the entire optimization problem reaches 1.5 h, which is much larger than the implementation period of the energy-saving strategy, and there is no possibility of implementation. Therefore, the community division of BS clusters is of great significance for reducing the computational time complexity.

In order to solve the problem of large-scale cellular networks and energy efficiency of energy-saving strategies, this chapter adopts the method of community division of BS clusters to realize the distributed implementation of energy-saving strategies. The construction of the base station network and the edges status are described in 3.1. The Louvain algorithm was used to divide the BS network into communities. The results of the community division are shown as follows: The number of nodes is 1444. The number of edges is 3821.The number of communities is 39. The maximum number of BSs included in the community is 107. The minimal number of BS included in the community is 2. The average number of BSs included in the community is 37. The modularity is 0.8745.

The total number of nodes of the BS in the network is 1444, and the total number of edges of the network is 3821, which is divided into 39 independent communities. The modularity range is $[-0.5, 1)$. The greater the degree of modularity is, the stronger the community characteristics of the networks and the greater the independence of the divided communities is. That is, the better the result of community division is. Studies have shown that when the value of modularity is greater than 0.5, the results of community division are better. The modularity of the BS network

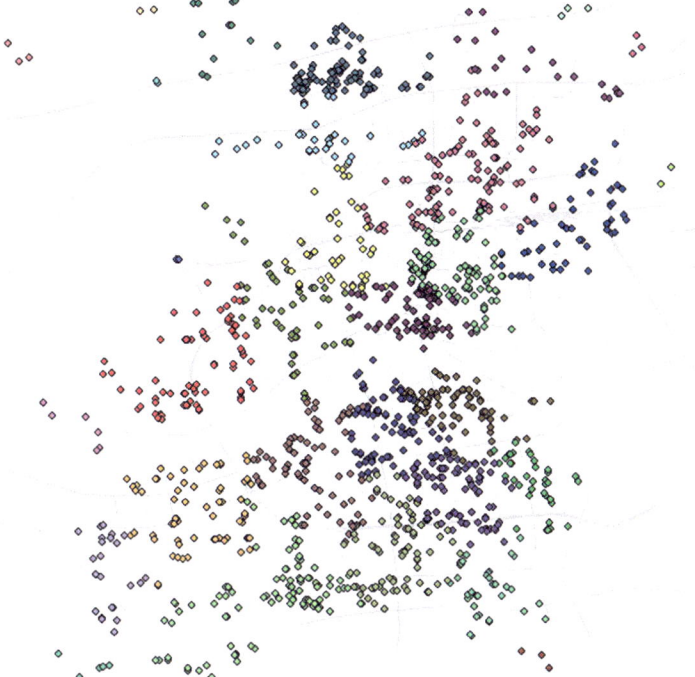

Fig. 4.15 Network community structure of base station

is as high as 0.8745. It shows that the BS clusters in cities have strong community characteristics and strong independence between communities. At the same time, in the divided communities, there is no situation where the number of BSs in the community is particularly large, and the community is divided more evenly, which facilitates the implementation of distributed computing.

With ArcGIS you can project points in space into real geographic space. Figure 4.15 projects the nodes of 1444 BSs in Lucheng District of Jinhua City onto the streets of Jinhua City. Different colors are used to mark different communities so that the distribution of sector nodes in the network and the community structure of the network can be seen. It can be seen that the nodes of different communities are separated in geographical space. The BSs in the same community are very close. The characteristics of the community structure are obvious, and the number of nodes in a single community is relatively uniform, which is conducive to the distributed implementation of the energy-saving strategy.

In order to verify the performance difference between implementing the BS energy-saving strategy for the BS cluster of the entire cellular network and implementing the energy-saving strategy for each community after dividing the community, this chapter chooses to implement ES-MSUE as a distributed implementation solution for two geographically adjacent communities, respectively. We combine two communities as a community to implement ES-MSUE as a centralized implementation plan and compare the results of the two scenarios. The number of cells selected for the community 1 is 54, and the number of cells selected for the community 2 is 59. The total number of cells after the merger is 113.

From Fig. 4.14, we can see that the complexity of solving the optimization problem has an exponential growth relationship with the number of variables. The experimental results show that the number of sectors in community 1 is 54. And the average calculation time is 8.14 s. The number of sectors in community 2 is 59. And the average calculation time is 23.7 s. But when merging the adjacent communities in these two spaces into a community with 113 sectors, the optimization solution time increases to 327 s, which is more than ten times.

We partition a day into 24 hourly intervals, the experiment is conducted on the off-peak hours from 0 o'clock to 6 o'clock. We first statistic the user numbers, access service types and traffic loads of cells. Then we construct the energy consumption model and user experience model according to the user-cell affinity graph. The object optimization problem is formulated and we solve this problem using branch-and-bound algorithm. Our strategy is compared with the energy-saving strategy proposed in [48] that only considered the flow-level performance (ES-FLD) and the classical energy-saving strategy that used the blocking rate as the user experience in literature [49]. The energy-saving strategy is compared with the aspects of the number of active cells, energy-saving ratio, and user QoE loss. And the result is shown in Fig. 4.16. From the experimental results, it can be seen that the energy-saving strategy ES-FLD achieves a maximum of 86% energy saving. It only considers the flow-level performance. The QoE loss increases several times, which seriously affects the user experience. DMUES proposed in this chapter achieves a

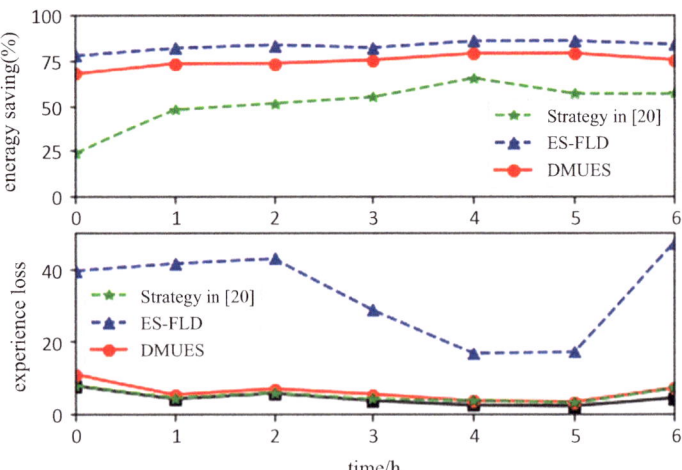

Fig. 4.16 Method performance at non-peak times

maximum of 79% energy saving, with very little increase in user experience loss with better performance.

Therefore, compared with ES-FLD, which considers only the flow-level performance cost, the DMUES achieves a higher energy-saving rate with only a little increase in the user experience cost. Compared with the energy-saving strategy in [19], under the premise of guaranteeing the cost of user experience, more than 20% of energy is saved. In general, compared with traditional base station energy-saving strategies, MUES achieves a good energy-saving effect and guarantees the quality of user experience, and achieves the best combination of the two.

At the same time, this chapter discusses the load utilization of active cells before and after implementing the DMUES at 0 o'clock, which is shown in Fig. 4.17. The results show that most of the cells in the city are under low utilization at 0 o'clock. And after the MUES is implemented, most of the cells with low utilization are turned off, while cells whose utilizations are high are basically been retained. The strategy also ensures that load of a single cell is in the affordable range. In conclusion, the strategy not only improves the energy efficiency of cellular network base station, but also ensures the load balance of the active cell. Furthermore, the computational complexity and time complexity are greatly reduced. This proves that the DMUES proposed in this chapter is very effective and feasible.

Current base station energy-saving strategy has not fully considered the impact of service characteristics on user experience and the problem of high time complexity of implementing the energy-saving strategy caused by the large scale of the 5G mobile network. In this chapter, we introduce user experience quality as the measure of the subjective experience of users on the received service and

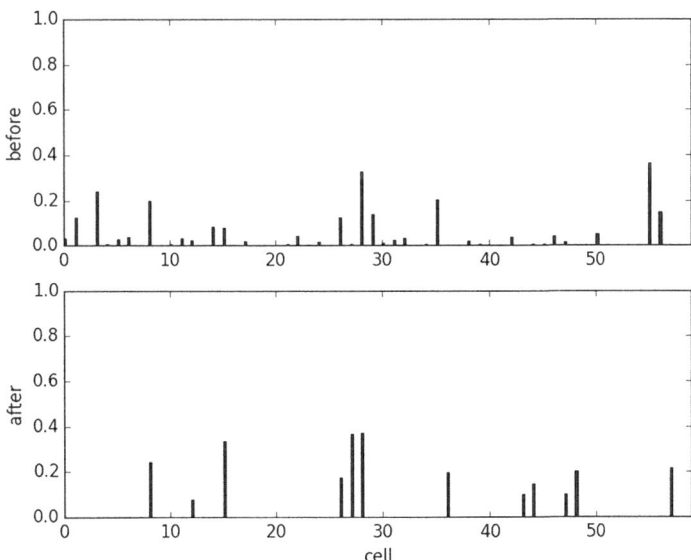

Fig. 4.17 Traffic loads of cells before and after implementing the DMUES

propose a distributed multiple-service user-experience energy-saving algorithm for base station. And we use a complex network graph model to describe the cooperation relationships between BSs based on the proximity of BSs and the closeness of interaction in the city. Using the method of community division, the urban BS is divided into multiple communities, and the energy-saving strategy is implemented separately for each community. Part of the cellular network in a city in China is selected, and the experiment was conducted on the non-peak hours in this scenario based on real UDRs. Compared with the existing energy-saving strategies, the results show that the DMUES proposed in this chapter can achieve good energy savings under the premise of less user experience loss. Furthermore, the computational complexity and time complexity are greatly reduced.

$$
\text{Hit}(Q) = \sum_{i=1}^{N} \sum_{j=1}^{M} \left(P_{ij} \cdot Q_{ij} + P_{ij} \cdot \left(1 - Q_{ij}\right) \cdot \right.
$$

$$
\left. \left(1 - \prod_{k \in \varphi^i, k \neq i} \left(1 - Q_{kj}\right)\right)\right)
$$

$$
\text{s.t.} \sum_{j=1}^{M} Q_{ij} S_j \leq V_i, i \in [1, N] \tag{4.59}
$$

4.3   Mobile Data Application in Sensing Networks

Over the past decade, the hotspots of wireless sensor network research have been gradually advanced from static networks deployed in specific monitoring areas to the perception of people's surroundings. Mobile perception is designed to take advantage of the mobility of vehicles, aircraft, ships, people, etc., combined with ubiquitous perceptual technology, to provide low-cost, more flexible perceptual data collection for monitoring objects away from network infrastructure through short-range wireless communications. In recent years, with the maturity of perceptual technology and application environment, mobile sensing networks are mainly committed to providing people with effective monitoring of the surrounding urban environment, providing a "human-centered" way of perception, and relying on human mobility. Understanding "human mobility and behavior preference" is very important for perceptual network applications, and this chapter will be elaborated from two aspects: human mobility analysis and preference pattern discovery.

4.3.1 *Human Mobility Analysis*

Human mobility has been widely investigated due to its application to a wide range of research fields, for example, urban planning (Barabasi [50], Becker et al. [51], Liu et al. [52]), traffic forecasting (Becker et al. [53], Peng et al. [54]), and epidemiology (Guo [55], Belik et al. [56], Liang et al. [57]). The analysis of mobility patterns has great significance to traffic modeling for simulation, forecasting, and control [58]. In addition, the identification of hotspot areas with unusually high levels of collective mobility is helpful for anomaly detection [59]. Moreover, the better understanding of human mobility is essential in epidemic prevention and control by limiting contacts with infected people (Longini et al. [60]). In the field of mobile communication, the spatiotemporal characteristics of human movement behavior can help mobile communication providers to estimate the traffic demands of users and make strategic decisions on resource allocation and service quality improvement.

For the past several decades, studies on human mobility patterns have primarily been based on census or questionnaire data, which are always accompanied with tremendous time and financial costs. In addition, due to the limited spatiotemporal coverage, methods based on such data fail to model human mobility accurately [61]. Thanks to the widespread use of location-aware devices, such as mobile phones and GPS-enabled devices, unprecedented amounts of records with high accuracy involving individuals trajectories have become available, providing an abundant data resource for human mobility research. Users usage detail records (UDRs) from cellular networks record spatial and temporal information when users access the base stations (BSs) for data usage. Since data usage behaviors are now very common in daily life and these records are generated spontaneously by users, it can accurately reflect human mobility behavior.

Human movement apparently exhibits temporal and spatial regularity, especially for large populations [52]. At the individual level, the statistical characteristics of individual movements [62–66] and the similarity of trajectories between individuals [67–70] have been investigated. At the aggregate level, research has been conducted from multiple perspectives, such as urban functional structure [71, 72], human activity [73], and community partition [74–76]. Furthermore, complex networks have been introduced into many geographical studies [77]. The derived spatial networks open up new avenues for human mobility analyses at the collective level. In these studies, city areas are transformed into nodes in spatial networks and spatial interactions between city areas are represented by weighted edges. Thus, the distance between nodes can be measured, and geographical information can be integrated into human mobility analyses. However, these studies, which are based on spatial networks, merely analyze spatial network structure, and the underlying collective mobility patterns are not further discussed.

In this chapter, we propose a collective mobility discovery method based on community differences (CMDCD) to analyze human mobility patterns at the collective level. First, we constructed spatial networks based on users UDRs in different periods and partitioned the networks into several communities using a community detection algorithm. Second, we identified community structure differences by analyzing the row norm of the exclusive-or (XOR) similarity matrix. Finally, since the network structural differences are caused by the change of collective mobility patterns, groups with different mobility patterns can be discovered based on the community differences.

The remainder of this chapter is organized as follows: Sect. 4.2 gives an overview of the related work. Section 4.3 introduces the study area and dataset used for this research.

Human mobility patterns have drawn much attention in the areas of physics, geography, and computer science with the availability of multi-source trajectory data. With a geo-tagged dataset, we can extract the footprints of large volumes of individuals. Although the trajectory of one particular person is relatively stochastic, we can find underlying patterns when the number of trajectories increases. In the age of big data, the widespread use of location-awareness devices, such as mobile phones and GPS-enabled devices, has made it possible to collect large-scale individual trajectories to analyze human mobility.

In 2005, Barabsi [78] found that there is increasing evidence that the timing of many human activities follows non-Poisson distributions, characterized by bursts of rapidly occurring events separated by long periods of inactivity. Brockmann's research on human spatial mobility suggested that the distribution of traveling distance decays as a power law, indicating that the trajectories are reminiscent of scale-free random walks known as Lvy flights [79]. The spatial and temporal characteristics of human mobility have laid the theoretical foundation for the study of a variety of human movement modes. Based on the dynamic analyses of human temporal and spatial mobility in these two articles, many related studies have emerged. These studies suggested that the spatial and temporal characteristics of human mobility have scale-free properties. At the individual level, based on the social media check-in data, Cheng et al. [63] found that the spatial movement distance between the users' consecutive check-ins and the radius of the user spatial mobility both follow power-law distributions. Gonzlez et al. [62] and Liu et al. [66], based on 10 million users' mobile communication data and 1.5 million taxi passengers' GPS data, respectively, found that the movement distance approximately follows an exponential truncated power-law distribution. At the aggregate level, Pei et al. [72] identified land use types in Singapore based on aggregated mobile phone data. Yuan et al. [71] discovered regions of different functions in a city using human mobility and points of interests (POIs). Jiang et al. [73] found that the population can be clustered into several representative groups based on different activity types. Taking advantage of check-in records, Sui et al. [76] constructed the network based on users trajectories between cities and found that the divided area boundary coincides with the administrative boundary. Kang et al. [75] explored human movements in Singapore based on taxicab usages data and found that trips within a community are much more than trips across different communities. In addition, many models have been proposed to explain human mobility patterns. These models consider different factors, such as population characteristics (Gonzlez et al. [62]), individual mobility (Song et al. [80]), geographical environments (Barabasi [50], Becker et al. [65]), and the distance effects (Liu et al. [66], Becker et al. [51], Zhou et al. [81]).

Recently, spatial networks have been increasingly introduced into the analyses of interaction patterns of urban space and collective mobility. In these studies, the regions of space are considered as nodes in the network, and the collective spatial interactions are considered as edges with different weights. These network nodes with spatial location information can reflect the impact that geospatial factors have on human mobility. By dividing the network space, the spatial area can be divided into different communities based on the contrastive analysis of urban community structure and administrative division. The relationship between different urban areas and the collective mobility patterns between different regions can be analyzed. Liu et al. [52] extracted nationwide interurban movements in China from a check-in dataset that covers half a million individuals and 370 cities to analyze the underlying patterns of trips and spatial interactions. By fitting the gravity model, they found that the observed spatial interactions are governed by a power law distance decay effect. They also constructed a spatial network where the edges denote the spatial interactions. The communities detected from the network are spatially connected and approximately consistent with province boundaries. De Montis et al. [82] found a similar phenomenon in the analysis of users commuter networks. Based on the mobile data of 100 million users, Gao et al. [83] attempted to explore and interpret patterns embedded in the network of phone-call interactions and the network of phone-users movements. They discovered high correlations between phone-users movements in physical space and phone-call interactions in cyberspace.

Our study area, Anshun, is a western city in China, consisting of six administrative districts. The dataset used for this research has been acquired from a Chinese mobile company, which is composed of users UDRs. The dataset based on UDRs contains data access records of 1.2 million mobile phone users over 21 days, covering 3980 BSs. The size of our dataset is shown in Table 4.4, and the spatial distribution of BSs is shown in Fig. 4.18, where dots with different colors represent BSs located in different administrative districts.

Spatial big data involves a large amount of data reflecting individual movement trajectories. Users UDRs from cellular networks record spatial and temporal information when users access the BSs for data usage, using the locations of

Table 4.4 Size of dataset

Number of records	275,465,112
Number of users	1,233,026
Number of BSs	3980
Time range of records	08/02/2013 00:00:00 to 28/02/2013 23:59:59

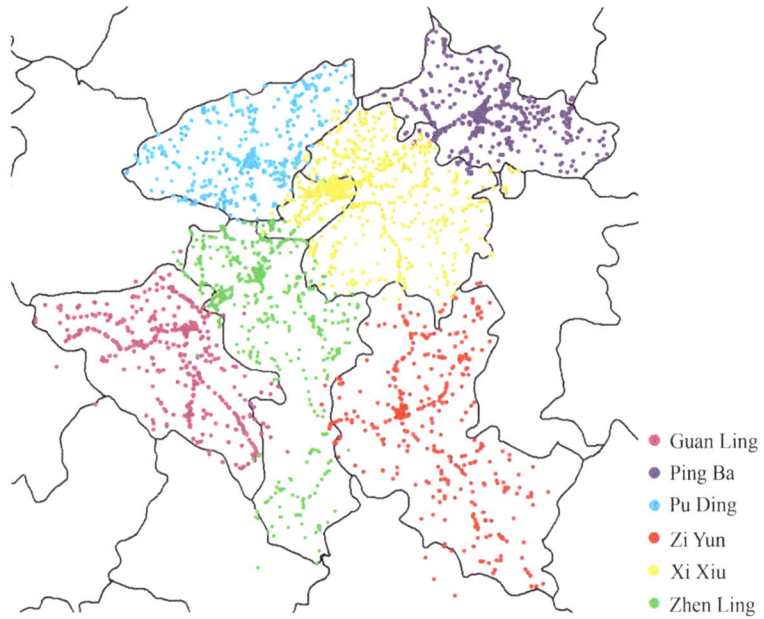

Fig. 4.18 Spatial distribution of base stations

Table 4.5 Description of UDR fields

Fields	Description
uid	An encrypted telephone number uniquely indicating a mobile user
stime	The time that a user begins to access the BS for data usage
longitude and latitude	Longitude and latitude corresponding to the BS's geographical location

BSs to represent users locations in geographic spaces. Compared with previous studies based on call detail records (Gonzlez et al. [62], Song et al. [80]), taking advantage of the UDRs, we extended the tracking of human movement from mobile call behavior to data usage behavior. Since data usage behaviors are now very common in daily life, UDRs provide a detailed description of human movement. A UDR includes the user ID, the timestamp of data usage behavior, and the location (longitude and latitude) of the accessed BS (Table 4.5). All the user IDs are anonymized for privacy protection. In cellular networks, a user always accesses the nearest BS, resulting in a positional data accuracy of approximately 300–500 m (Yuan et al. [84]). Therefore, each users trajectory can be approximated by the location sequence of the BSs (Fig. 4.19).

The records were filtered using the following steps. To eliminate the abnormal records caused by the oscillation effect (Wu et al. [85]), we identified abnormal

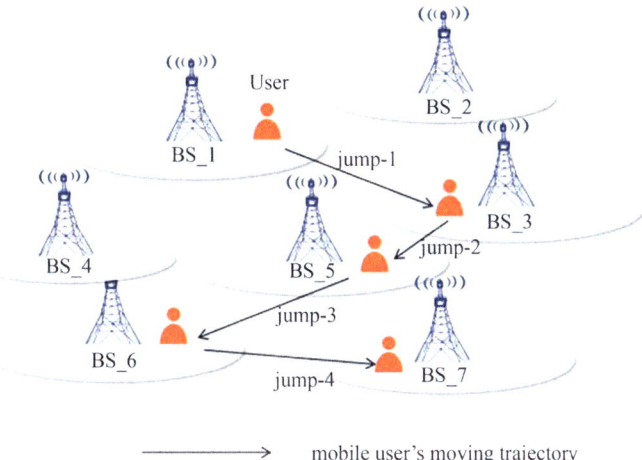

Fig. 4.19 User's moving trajectory based on UDRs

records by calculating the moving speed between adjacent BSs. We excluded abnormal records when the movement speed exceeds the threshold of 120 km/h as suggested in Wang et al. [86]. In addition, individuals who visit only one BS are considered stationary, and such records were removed.

Furthermore, the purpose of this study is to analyze collective mobility patterns in different periods. Human movements present significantly different patterns between workdays and holidays. For example, we usually have more regular routines in workdays due, with commuting behavior between home and workplace representing a large proportion of daily trips. In contrast, during holidays, people may go out for entertainment or traveling, thus resulting in more random trajectories. Therefore, the Spring Festival vacation and workdays are taken as examples to analyze the changes in user movement behavior. The records of 4 days during 09/02/2013–12/02/2013 (the Spring Festival vacation, namely the Chinese New Year; this period is referred to as H) and during 18/02/2013–21/02/2013 (workdays; this period is referred to as W) were selected. To explore the statistical properties of human movement behavior in period H and period W, we calculated the cumulative distribution function of movement frequency and movement radius (Gonzlez et al. [62]) (Fig. 4.20). More than 85% of users have a movement frequency less than 25 in period H, while the proportion in period W is only 69.6%. The proportions of users whose movement radius is within 5 km in period H and period W are, respectively, 81.3% and 62.8%. Users' movement frequency in period W is much higher than that in period H, and the users' movement area covers a wider range in period W. The statistical results indicate that the human mobility intensity is enhanced in period W.

Given the preceding analysis, 2000 users were selected in each administrative district for the large-scale analysis. For the finer scale analysis, each administrative district was further divided into 25 km by 25 km grids. Reliable and representative

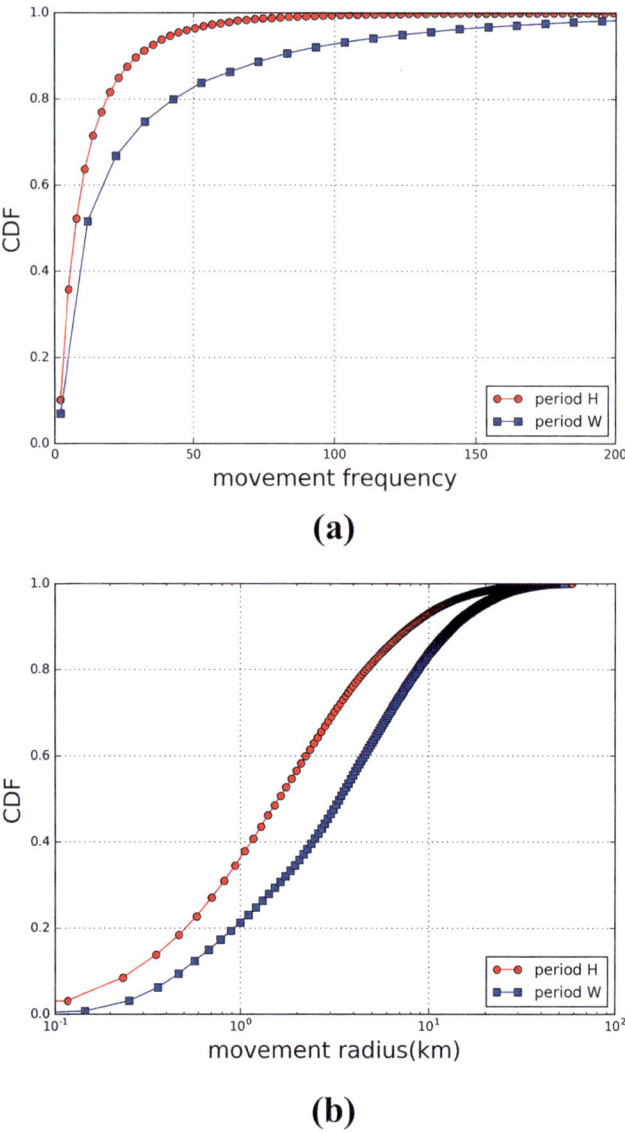

Fig. 4.20 Cumulative distribution function (CDF) of (**a**) movement frequency and (**b**) movement radius

candidate users were defined as those that met the following criteria. In period H, each user was active only within a single grid. In period W, 900 users moving between Guanling and Zhenning are referred to as $U_{\text{Large\_scale}}$, while 600 users moving across grids in Guanling and Zhenning are referred to as $U_{\text{Small\_scale}}$. The $U_{\text{Large\_scale}}$ and $U_{\text{Small\_scale}}$, respectively, represent users whose mobility patterns change at the large spatial scale and small spatial scale. In the four other administrative districts, the users remained active within a grid in which they were located in period H. The filtered dataset for our analysis contains 2.01 million records from 12,000 users. Finally, each record is represented as a tuple of ⟨ uid, stime, longitude, latitude ⟩. Each user's trajectory was constructed by appending all the recorded locations with the same uid in chronological order based on the timestamps.

Spatial networks are introduced to analyze human mobility patterns at the collective level. We constructed spatial networks in which nodes represent BSs, and edge weights denote collective mobility intensity between BSs. The change of collective mobility patterns affects connection relations between nodes in networks, thereby leading to network community structural change. We partitioned the networks into several communities to analyze the community structures and identified community differences. We can further discover users whose mobility patterns have changed based on community differences.

In this chapter, we propose a CMDCD method to uncover users whose mobility patterns have changed in different periods. The method comprises three core steps (Fig. 4.21). The first step is to construct spatial networks based on users' UDRs in different periods and partitioned the networks into several communities using a community detection algorithm based on stability. The second step is to identify community structure differences by analyzing the row norm of the XOR similarity matrix. Third, users with different mobility patterns can be discovered based on the community differences. This method can be applied to analyze collective behavior in different periods. By applying the method to data UDRs collected from the cellular networks in Anshun, we analyzed different collective mobility patterns in the Spring Festival vacation and workdays.

With advances in complex network research, many geographical studies have introduced spatial networks into geographical analyses. Spatial networks are widely used; each node is located in space so that the distance between each pair of nodes can be measured (Wang et al. [77], De Montis et al. [87]). Spatial networks are networks for which the nodes are located in a space with a metric. A spatial network may be tangible (e.g., street networks) or intangible (e.g., flight networks or networks constructed from social media), where the edges denote the spatial interactions. Based on methodologies for building the network, current typical network models mainly include interaction networks (Gao et al. [83]) and collaboration networks (Newman [88], Wu et al. [89]).

Figure 4.22 shows the effect of changes in the user's trajectory on community structures in these two types of networks. Dots with different colors in Fig. 4.22 represent BSs belonging to different communities, black lines represent edges which the user contributes to the network in the current period i, and red lines represent edges which the user contributes to the network when the user's trajectory changes

Fig. 4.21 Framework of the
CMDCD method

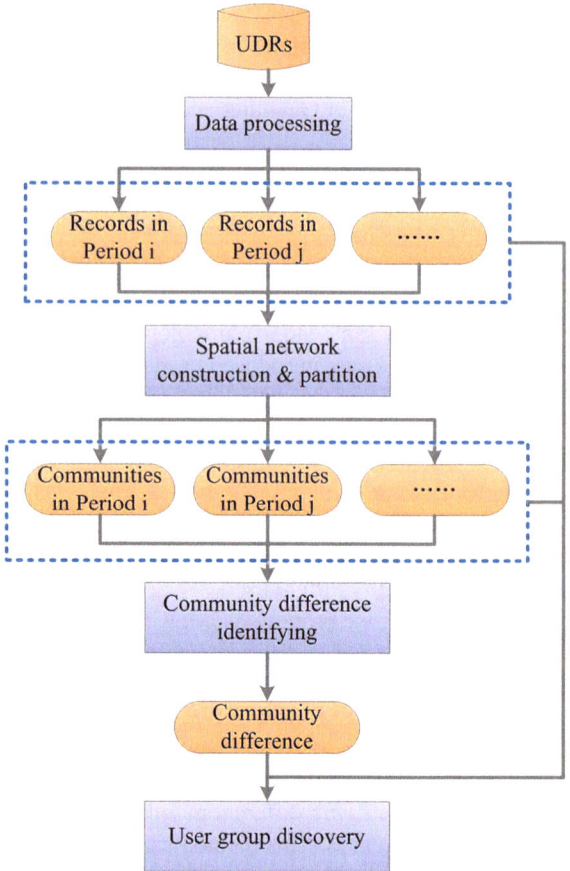

(the BSs the user accessed increase) in period j. The users mobile trajectory in
period i is $A \rightarrow B \rightarrow C \rightarrow D$ and crosses two communities C1 and C2. In period
j, the user additionally accesses BS E.

1. If the user's mobile trajectory is $E \rightarrow A \rightarrow B \rightarrow C \rightarrow D$: in the interaction
   network, an edge (E,A) is added between E and C1, and E tends to be partitioned
   into C1; in the collaboration network, three edges (A,E), (B,E), and (C,E) are
   added between E and C1, and an edge (D,E) is added between E and C2. The
   connection relation between E and C1 is stronger than that between E and C2, so
   E tends to be partitioned into C1.

2. If the user's mobile trajectory is $A \rightarrow B \rightarrow C \rightarrow D \rightarrow E$: in the interaction
   network, an edge (E,A) is added between E and C2, and E tends to be partitioned
   into C2; in the collaboration network, three edges (A,E), (B,E), and (C,E) are
   added between E and C1, and an edge (D,E) is added between E and C2; thus, E
   still tends to be partitioned into C1.

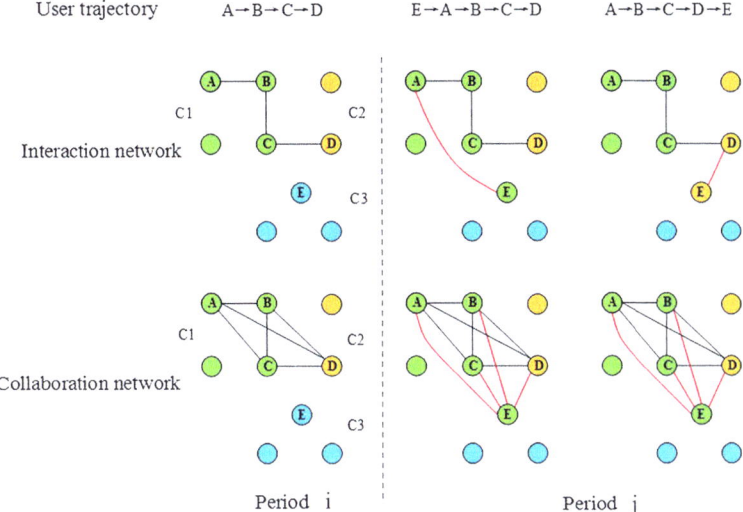

Fig. 4.22 Effects of user trajectory change on the network structure

When users collectively move from one activity area to another, the community structures in the interaction networks can intuitively reflect this collective mobility behavior. Because there is no scheduling in the network model building of the collaboration network, its network structure has a memory effect on the perception of users mobility behavior, and changes in the community structure are related to not only users current mobility behavior but also users historical moving trajectories, which cannot directly reflect changes in collective mobility behavior. Therefore, the interaction network is chosen to analyze the collective mobility behavior in this chapter.

In geographic information science, nodes of spatial network are used to characterize the spatial region or spatial entities, while weighted edges represent the intensity of connections between nodes. Taking advantage of UDRs, the location sequence of BSs a user accessed can be used to approximate the users movement trajectory. In order to aggregate and analyze movements of individuals at the collective level, geo-tagged BSs are transformed into nodes, while interaction relations embedded in space are represented by weighted edges.

In this chapter, based on UDRs, we can construct spatial interaction networks where nodes represent the BS, and weighted edges measure the interaction between BSs formed in the process of providing data services to the same group of mobile users. If two BSs are visited by the same user in succession, there will be an edge between them in the spatial interaction network. The weight of the edge is defined as

$$\omega_{ij} = \sum_u \delta^u_{ij} \qquad (4.60)$$

where δ_{ij}^u is 1 if BS i and BS j are visited by mobile user u in succession and 0 otherwise. The physical meanings presented in the spatial interaction network are that nodes contain spatial information and that edges are generated by integrating human movement.

A community detection algorithm based on stability was applied to partition the network into several subnets [90], see Schaub et al. [91]. It adopts Markov dynamics as a zooming lens for multiscale community detection. Compared with other algorithms (e.g., modularity) affected by a resolution limit, a lower scale that establishes a minimum size below which communities cannot be detected, the algorithm based on stability uses the evolution of a Markov process on the graph as a zooming lens over the structure of the network at all scales. The Markov time also acts affectively as an intrinsic resolution parameter that establishes a hierarchy of increasingly coarse communities.

This algorithm approximates community detection as a dynamic Markov process:

$$\dot{p} = -p[D^{-1}L] \tag{4.61}$$

where p is a $1 \times N$ probability vector, D is the diagonal matrix $D_{ij} = d_i = \sum_j A_{ij}, A$ A is the adjacency matrix, and $L = D - A$ is the Laplacian matrix.

The clustered autocovariance matrix of the network at time t is defined as

$$R_t = H^T \left[\prod^{\exp}(-tD^{-1}L) - \pi^T\pi \right] H \tag{4.62}$$

where $\prod \mathrm{diag}(\pi)$, the stationary distribution π of this dynamic is $p_i = D_{ii}/2m$, and partitioning the network into c communities is encoded into a $N \times c$ matrix H with $H_{ij} \in \{0, 1\}$, where 1 denotes that node i belongs to community j.

The stability of a partition H at time t is defined as

$$r(t, H) = \mathrm{trace}[R_t] \tag{4.63}$$

For each Markov time t, we seek the partition with the largest stability.

$$r(t) = \max Hr(t, H) \tag{4.64}$$

The existing index used to compare different clustering results, such as accuracy, recall, F-measure, MOC, and D-measure (Pfitzner et al. [92], Schieber et al. [93]), measures the clustering result difference as a whole and is unable to locate the definite partial difference. To compare the partial difference of community structure, existing research has mapped the community structure to geographic space and then observed and subjectively judged the different community structures. Clearly, the random error of this method increases and the reliability of the results cannot be

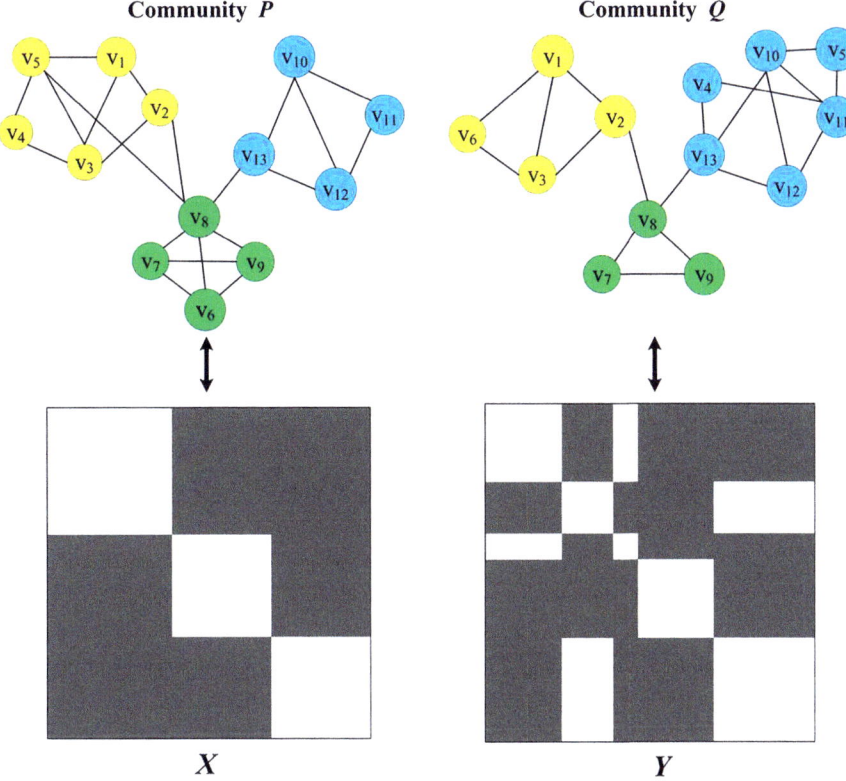

Fig. 4.23 Network community structure and similarity matrix

guaranteed. Therefore, this chapter identified community differences in different spatial networks by analyzing the row norm of the XOR similarity matrix.

Given two sets of communities, represented as $P = \{P_i\}(1 \leq i \leq m)$ and $Q = Q_j(1 \leq j \leq n)$, and the nodes in the network are denoted as $V = \{v_i\}(1 \leq i \leq N)$, the procedure for identifying community differences is as follows:

First, the similarity matrices X and Y are constructed for two sets of communities (Fig. 4.23). In the first set of communities, if the nodes v_i and v_j are partitioned into the same community, that is, $\{v_j \in P_k, v_j \in P_k\}$, the corresponding similarity X_{ij} in matrix X is 1. When the nodes v_i and v_j belong to different communities, that is, $v_i \in P_k, v_j \in P_k$, the similarity is 0. The same procedure may be easily adapted to obtain similarity matrix Y for the second set of communities. Then, we XOR matrix X with Y to construct XOR similarity matrix M, $M = X \oplus Y$. The white block in the matrix represents the similarity value of 1, and the gray block represents 0.

The row norm of the XOR similarity matrix M is calculated as

$$\| M_j \|_1 = \sum_j | M_{ij} | \tag{4.65}$$

The node v_i belongs to the community P_k and Q_j in the two sets of communities, respectively, $\{v_i \in P_k \mid P = \{P_i\}\}$ and $\{v_j \in Q_j \mid Q = \{Q_j\}\}$. The row norm of node v_i can be deduced from the number of nodes in community P_k and Q_j, denoted as $| P_k |$ and $| Q_l |$:

$$\| M_i \|_1 = | P_k \cap (U - Q_l) + Q_l \cap (U - P_k) | = | P_k | + | Q_l | - 2 \times | P_k \cap Q_l | \tag{4.66}$$

if $| P_k \cap Q_l | \geq \gamma | P_k |$ and $| P_k \cap Q_l | \geq \gamma | Q_l |$, $\{P_k \cap Q_l\}$ takes a large proportion of both P_k and Q_l. The differences are actually parts of $P_k - \{P_k \cap Q_l\}$ and $Q_l - \{P_k \cap Q_l\}$, and the corresponding community structure $\{P_k \cap Q_l\}$ is considered unchanged. Therefore, the node v_i is not different. While $| P_k \cap Q_l | < \gamma | P_k |$ or $| P_k \cap Q_l | < \gamma | Q_l |$, $\{P_k \cap Q_l\}$ belongs to smaller than γ parts of either P_k or Q_l, and the node v_i is identified as different between P and Q.

As shown in Fig. 4.24, community differences between the two sets of communities P and Q could be identified by analyzing the row norm $\| M_i \|_1$ of the XOR similarity matrix M In practice, when the row norm satisfies $\| M_i \|_1 > (1 - \gamma)$ $(| \overrightarrow{P} | + | \overrightarrow{Q} |)$, the node v_i is identified as the difference between two sets of communities, where $| \overrightarrow{P} |$ and $| \overrightarrow{Q} |$ represent the average number of nodes in each community of P and Q, respectively.

Because edge weights denote collective mobility intensity, changes of users' collective mobility patterns affect connection relations between nodes in networks, thereby affecting community structures. Based on differences in community structures, this chapter identified users whose mobility patterns change in different periods.

First, the row vector set $\{M_i\}$ corresponding to the community differences $\{v_i\}$ is found in the XOR similarity matrix M. Then, the nonzero elements in $\{M_i\}$ are

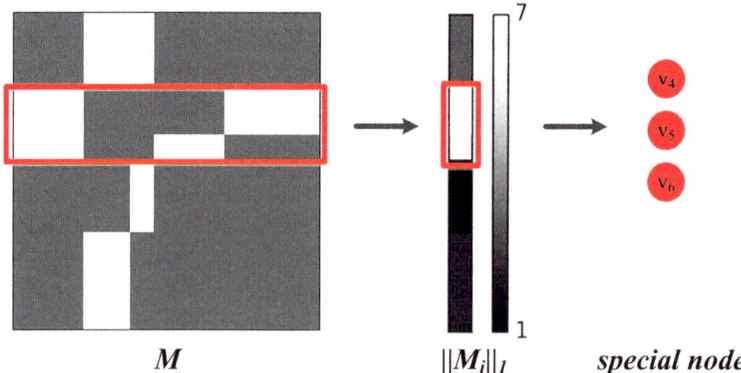

Fig. 4.24 Identification of community structure differences

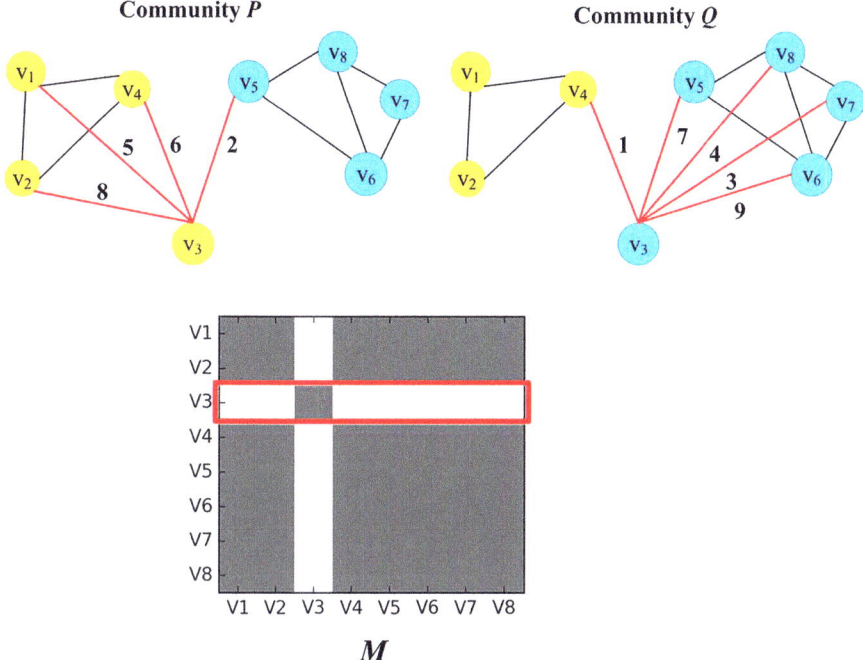

Fig. 4.25 Network community structures and XOR similarity matrix

found. If $M_{ij} \neq 0$, the corresponding edge (v_i, v_j) to the node v_i and v_j is the edge that has an impact on the network community structure, denoted as E_{differ}.

As shown in Fig. 4.25, dots with different colors represent nodes belonging to different communities and numbers beside the edges denote the edge weights. Since the edge weights of node v_3 have changed, the node v_3 is partitioned into different communities in two sets of communities. By analyzing the row norm, the node v_3 is identified as the difference between two sets of communities. The red lines in Fig. 4.25 represent edges that have an influence on the community structure, and the red box represents the row vector corresponding to the node v_3 in the XOR similarity matrix M. Through the nonzero element in the row vector, E_{differ} can be identified.

The edge weights in the spatial network reflect the intensity of the users mobility behavior, and the change of collective mobility patterns leads to network community structural change. Therefore, based on the edges influencing community structure, we can discover users whose mobility behaviors exhibit differences in different periods.

The edges which the user u_i contributes to the network are denoted as E_i; if there are edges that affect the community structure in E_i, that is, $E_i \cap E_{\text{differ}} \neq \emptyset$, the user u_i is identified as the user whose mobility patterns have changed in different periods.

The statistical results of the community detection are shown in Fig. 4.26. With increasing time, the Markov process explores larger regions of the network; therefore, the Markov time acts as a resolution parameter that enables us to identify community structure at different scales (Table 4.6).

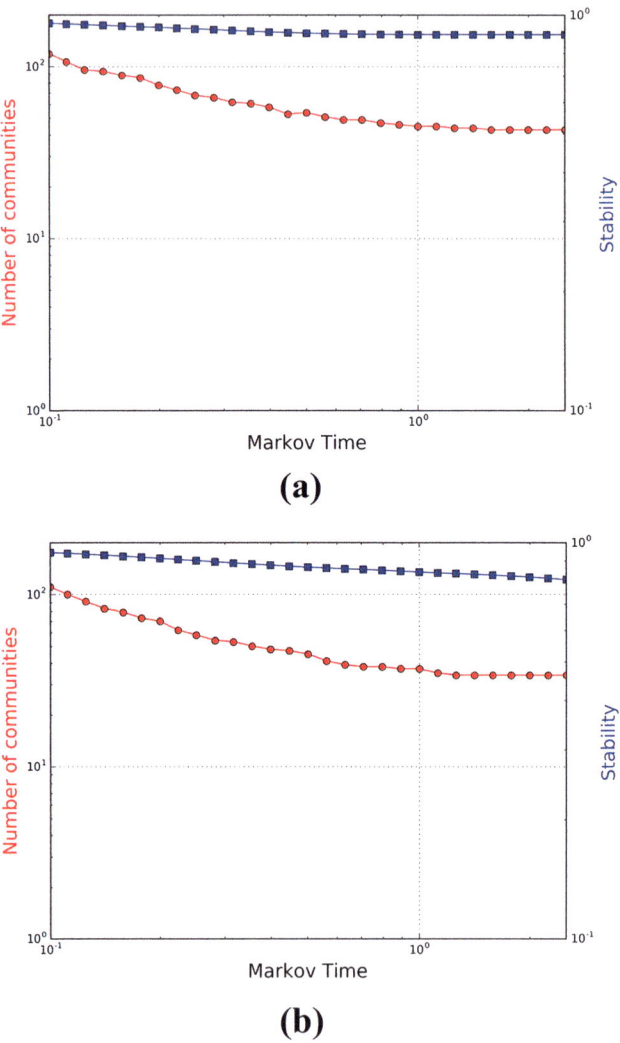

Fig. 4.26 Community detection results of spatial interaction networks in (**a**) period H and (**b**) period W

Table 4.6 Size of spatial interaction networks

Period	Number of nodes	Number of edges
H	1898	9980
W	1882	13,150

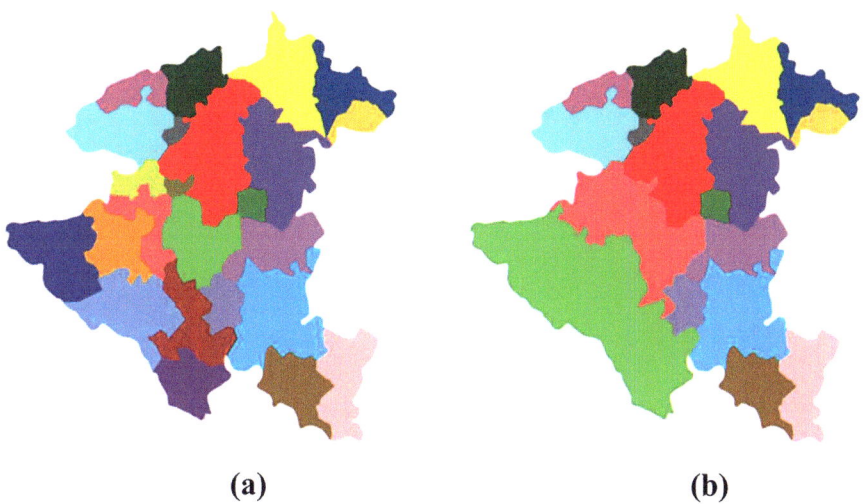

(a)   **(b)**

Fig. 4.27 Community structures of spatial interaction networks in (**a**) period H and (**b**) period W

Community structures were mapped to geographic space using ArcGIS and Google Earth. As can be seen in Fig. 4.27, there are more communities in Guanling and Zhenning during period H and the community scale is smaller compared with that in period W due to weak mobility intensity. Since the collective mobility intensity is enhanced in period W, the number of communities decreases and each community covers a wider range. The differences in community structures reflect the impact of different mobility patterns on the network structure.

Following the methods, community differences were identified in networks constructed in the two periods. In order to measure the difference in network community structures on the whole, we introduced the normalized mutual information (NMI), which is used to quantitatively express the coincidence between different clustering results (Mcdaid et al. [94]). Given two sets of communities P and Q, NMI is defined as

$$\text{NMI}(P, Q) = \frac{2 \times I(P, Q)}{H(P) + H(Q)} \tag{4.67}$$

where $I(P, Q)$ is the mutual information of P and Q, $H(P)$ and $H(Q)$ are the entropy of P and (Q), respectively. This provides a measure of the degree of coincidence between two clustering results P and Q, taking on a value of 0 for independence between P and Q and 1 when $P = Q$.

In Fig. 4.28, the blue line represents the NMI of two community structures in period H and period W, and red line represents the NMI of two community structures excluding these identified community differences. The NMI value is significantly improved after excluding the differences between communities in the two periods, indicating a greater coincidence between community structures. These imply that

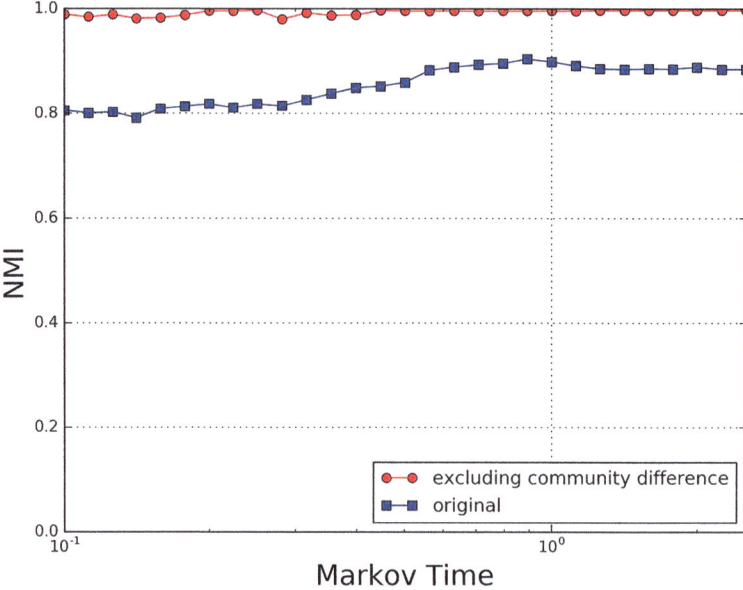

Fig. 4.28   The NMI of two community structures in period H and period W

the proposed method can effectively identify local differences between community structures.

According to the user group discovery method proposed, Fig. 4.29 shows that the CMDCD method achieved a precision of 94.1% and a recall of 86.7%. The method can discover 85.1% $U_{\text{Large\_scale}}$ and 89.1% $U_{\text{Small\_scale}}$ (4.29). The experimental results indicate that the CMDCD method proposed in this chapter can effectively identify users whose mobility patterns have changed in different periods at the large spatial scale and the small spatial scale.

Since the change of collective mobility patterns has an influence on community structure, the community structure differences will be reduced when the impact of users with different mobility patterns is eliminated. Therefore, the edge weights which the discovered $U_{\text{Large\_scale}}$ and $U_{\text{Small\_scale}}$ contributed to the spatial network in period W were replaced by those in period H. Figure 4.30 shows the community structure after eliminating the influence of the discovered users whose mobility patterns have changed in different periods. As can be seen, the updated community structure in period W and the community structure in period H are approximately the same. The NMI of these two community structures is 0.984, while the NMI of the original community structures in the two periods is 0.885. The experimental results demonstrate the effectiveness of the CMDCD method proposed in this chapter.

Furthermore, we compared the performances of the CMDCD method and the methods based on cosine similarity, Pearson's correlation coefficients, Kullback–Leibler divergence, and Jensen–Shannon divergence, respectively. Li et al. [67] utilized cosine similarity and Pearson's correlation coefficients to mine user sim-

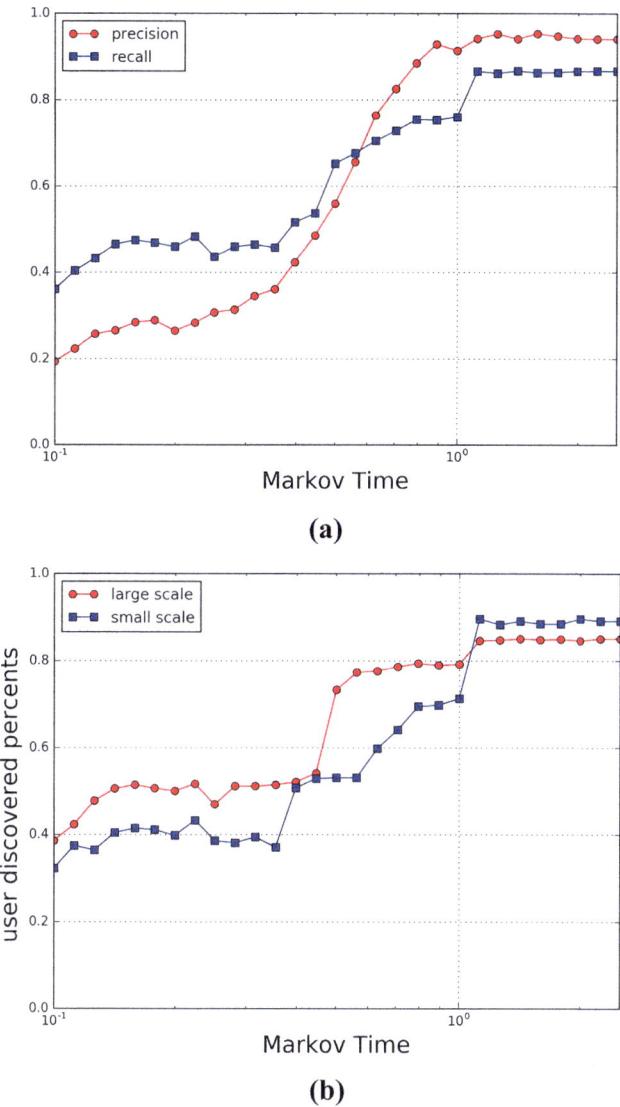

Fig. 4.29 The performances of the CMDCD method. (**a**) The precision and recall. (**b**) The $U_{\text{Large\_scale}}$ and $U_{\text{Small\_scale}}$ discovered proportion

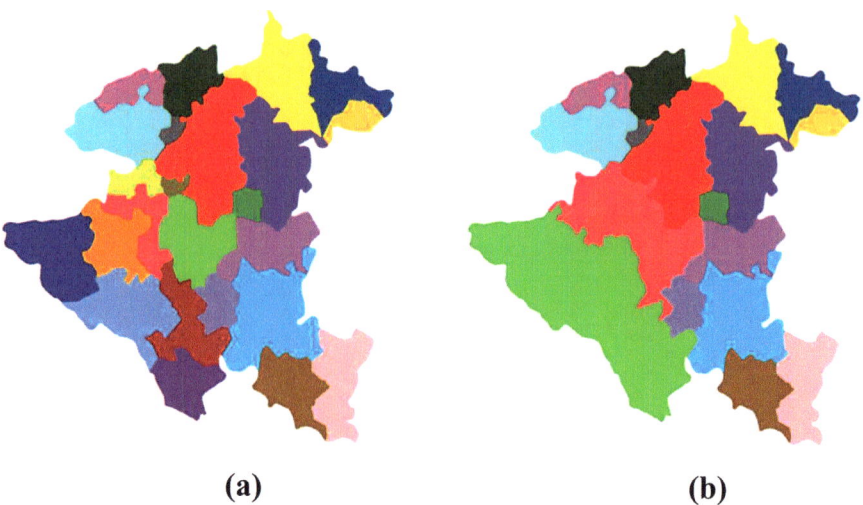

Fig. 4.30 Community structure in (**a**) period H and updated community structure in (**b**) period W

ilarity based on location history. Lv et al. [70] analyzed the similarity of visiting patterns by calculating Kullback–Leibler divergence. These measures have been claimed to outperform other existing similarity measures (Sarwar et al. [95]) and are widely used in recommendation systems. The CMDCD method proposed in this chapter outperforms the methods based on these measures in terms of precision, recall, the $U_{\text{Large\_scale}}$, and $U_{\text{Small\_scale}}$ discovered proportion (Fig. 4.31). Our numerical findings above show that our approach presents an outstanding combined qualitative–quantitative framework to identify users whose mobility patterns have changed in different periods.

Exploring collective mobility and distinguishing mobility patterns between different periods are highly important for a variety of domains such as urban planning, epidemiology, traffic management, and resource allocation. Many previous studies have introduced spatial networks into human mobility analyses at the collective level. However, these studies merely analyzed spatial network structure, and the underlying collective mobility patterns are not further discussed. In this study, we proposed a CMDCD to analyze human mobility patterns at the collective level.

Through introducing the spatial network and integrating collective mobility intensity into weighted edges, the method discovered users whose mobility patterns have changed based on community differences. The results indicate that the CMDCD method has achieved the goals of identifying community differences and discovering users with different mobility patterns simultaneously. Our finding agrees with previous studies (Gao et al. [83]) that the change of collective mobility patterns affects the connection relations between nodes in networks, resulting in community structure differences. We further discovered users whose mobility patterns exhibit differences based on community differences. The results achieved

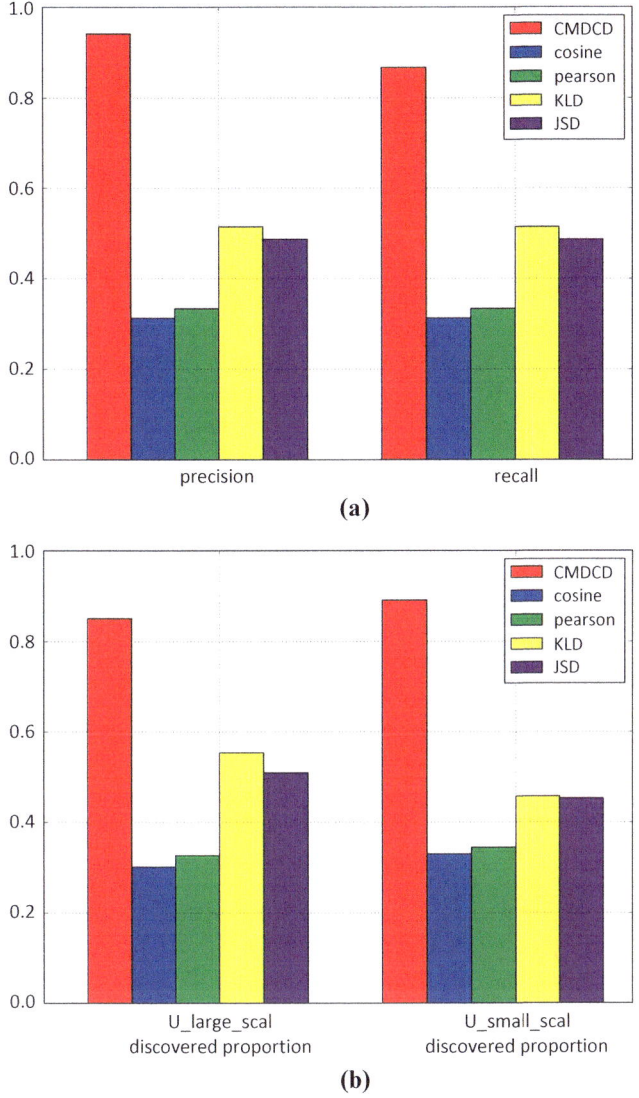

Fig. 4.31 The performances of typical measures compared with the CMDCD method. (**a**) The precision and recall. (**b**) The $U_{\text{Large\_scale}}$ and $U_{\text{Small\_scale}}$ discovered proportion

a precision of 94.1% and a recall of 86.7%. Moreover, the CMDCD method can discover 85.1% U_{Large_scale} and 89.1% U_{Small_scale} indicating that the method is effective for the discovery of groups whose mobility patterns have changed at different spatial scales.

The CMDCD method is effective in distinguishing human mobility patterns in different periods. Furthermore, it can handle a variety of location-based datasets by introducing spatial networks that is of considerable significance. On the one hand, human mobility patterns at different spatial scales can be investigated. On the other hand, various information can be integrated into human mobility analyses, such as the activity information provided by check-in records and the text attributes of query logs, which broadens our horizons of analyzing human mobility patterns. We can also investigate the underlying motivations driving human movements combined with geographical context and socioeconomic factors.

As shown in Fig. 4.32, areas ① and ② are partitioned into different communities in the Spring Festival, and in workdays, they are aggregated into a single community. According to the geographical context, we know that area ① is a town of Anshun named Huangguoshu, which contains a famous 5A level scenic spot, Huangguoshu Waterfall Scenic Area, and area ② is located near the highway G60. During the Spring Festival, most people celebrate it with their family members at home, leading to weak mobility intensity. In workdays, however, the scenic area will attract large numbers of tourists, and the highway G60 near the scenic area is the main channel for tourists, so the number of users moving between the two regions increases dramatically, and the spatial interactions between BSs of the two communities become stronger. The statistical analysis also demonstrates that collective mobility

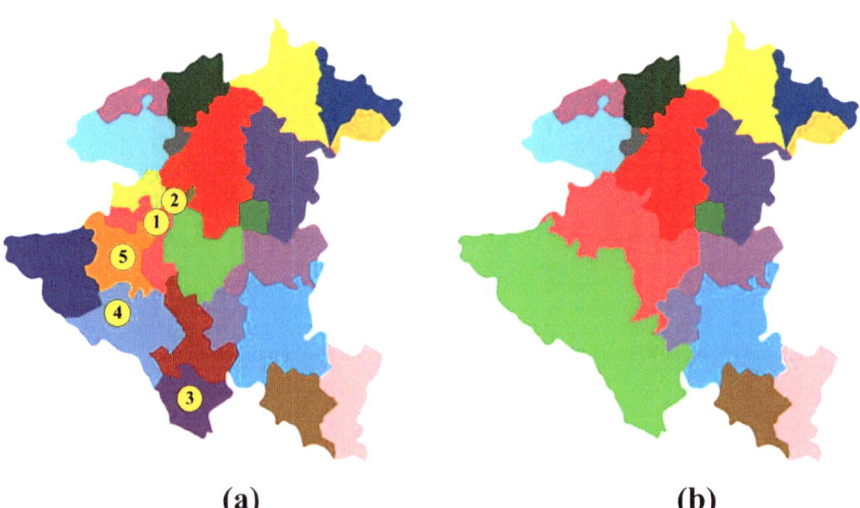

(a)   **(b)**

Fig. 4.32 Areas around scenic spots and highways in community structures in (**a**) period H and (**b**) period W

intensity is enhanced in workdays. Furthermore, areas ① and ③ belong to the same prefecture, but they are partitioned into different communities in both periods. Since there are mountains in these areas, and there are no roads directly connecting them, there are fewer movements between ① and ③ and they are thus separated. However, highway S210 is between ③ and ④ which directly connects them, and highway S50 connects ③ and ⑤, enabling easier human movements. Therefore, areas ③, ④, and ⑤ are aggregated into a single community during workdays. This indicates that human mobility is influenced by socioeconomic factors, urban road network, and geographical barriers (e.g., mountains and rivers).

In the era of big data, along with large-scale geo-tagged data, analysis of human mobility has become a hot research topic in many fields, such as urban planning, traffic forecasting, epidemiology, and business recommendation systems. This chapter analyzed collective mobility patterns by identifying community differences in spatial networks.

In this chapter, we proposed a CMDCD method to analyze human mobility patterns at the collective level. First, we identified the community differences between spatial networks, and then we discovered users with different mobility patterns based on community differences.

We validated the CMDCD method using empirical UDRs collected from the cellular networks in a city in China. The results of the experiment have demonstrated the effectiveness of the method. Experimental results show that the CMDCD method can solve these two problems of identifying community differences and discovering users with different mobility patterns simultaneously. Furthermore, the CMDCD method is an integrated approach to discover groups whose mobility patterns have changed in different periods at the large spatial scale and the small spatial scale.

The research in this chapter has great importance for the construction of smart cities and also provides a new view for analyzing spatial interactions based on collective human mobility. A better understanding of collective mobility patterns in different periods can help manage traffic flows and plan public transportation services for smart transportation. Additionally, exploring human mobility patterns enables urban planners to make strategic decisions on land use, urban planning, and business investments for smart economies. Moreover, city authorities can identify the hotspot areas where collective mobility intensity exhibits great differences compared to normal periods for public safety, which is important for crowd management and anomaly detection.

4.3.2   Preference Pattern Discovery

Mobile phones have become the necessities of human life, and bring kinds of necessary services in social life. In the year of 2016, the sales of smartphones around the world reached 1 billion and 470 million. Mobile applications are the platform for user network life. As of March 2017, the number of Google Play applications has reached 2 million 800 thousand, and the number of App Store applications has reached 2 million 200 thousand. The mobile phones have become users' access to the Internet, so understanding the relationship between mobile phone and users' Internet behavior mode is very important for service providers and mobile operators' recommendation and market strategy formulation.

In this background, mobile phone manufacturers and Internet service providers urgently need to understand the user's requirements and Internet behavior patterns when using different mobile phone brands. On the one hand, by finding out the relationship between APP and brand in user's Internet behavior, mobile manufacturers can improve user experience by preloading APP. On the other hand, service providers can make effective recommendation strategies through the relationship by analyzing the user's transfer behavior between different APP to attract more people to use their service. Therefore, finding the relationship between mobile phone brand and APP can produce huge economic benefits for both the mobile phone manufacturer and the service provider. To find out the characteristics of Internet behavior pattern, the traditional method is survey and telephone interview. This method usually consumes more resources, such as time, money, and so on. Another method is use user's private information. Previous studies have shown that users' Internet access patterns may be related to many potential factors, such as age, interest, and occupation. But user's private information are often not accessible to privacy. The user details record (UDR) accurately and anonymously records important information of mobile phone users' Internet activities through cellular network and is easy to get, so it became the main tool to understand individual behavior characteristics in the era of big data.

However, all types of mobile Internet behavior, such as engaging in Internet activities, doing online shopping, watching network videos, and so on, will generate large amounts of UDR data on the operator side. In such a large scale of data, mining all useful information is obviously hard and not practical. But we believe that there must be commonalities in user's Internet behavior although the Internet behavior is different and complex. So, digging out commonalities in different Internet behavior, called typical behavior patterns, from the big data of mobile users is a practical and efficient choice to describe user's mobile Internet behavior.

Therefore, it is our choice to use UDRs to analyze user's Internet activities and typical behavior patterns hidden in different behaviors. To find out the Internet behavior patterns efficiently and to provide business guidance for mobile phone producers and service providers reasonably, we put forward the concept of mobile Internet life persona (MILP). It is a kind of typical behavior patterns which is used to describe user's transfer behavior between different APP. According to it, we can know the behavior characteristics of users when switching APP, and then we could discovery the law of MILP and the relevance between MILP and time which provides more useful and reasonable business advice. On this basis, we need to understand:

• What is the relevance between different mobile phone brands and different APP?

- How to find mobile Internet life personas through UDRs, and what is the homogeneity and heterogeneity of mobile Internet life personas of different mobile phone brands?
- What is the relationship among the mobile phone brand, the user's Internet time, and the mobile Internet life personas?

Therefore, based on the above problems, we need an easy and practical method, which can directly display the Internet behavior patterns of users using different brand phones.

We provide a framework for mining mobile users' typical sequential patterns in Internet services, which mainly use thematic models and feature extraction ideas. This framework first analyzes the relationship between mobile phone brand and APP from the point of feature decomposition. Then the concept of MILP is introduced. It will reveal to us the characteristics of users switching between different APP. Latent MILP indexing thematic model is proposed to discover MILP and weights of MILP. Next, we will find the relationship between MILP and time that users use mobile phones. Through the above framework, we can analyze a large amount of UDRs data, transform it into discrete time series, and extract hidden themes and strong rules hidden in the data to design user portraits of different groups of people represented by different kinds of characteristics.

The rest of this article is set as follows: In the second section, we will summarize the current research status on user typical pattern mining. In the third section, the specific framework and process of the study are explained. The experimental results and discussions are in the fourth section. The fifth section summarizes the work and contribution of this article and looks forward to the further direction of future research.

Many researches have been done in user portrait analysis, and many effective methods have been put forward to characterize users' habits of Internet.

However, in the traditional methods of analysis, some people are only committed to analyzing the user's Internet habits from the social attributes of the user itself. For example, Luo et al. [96] analyze user's behavior through social network models based on the user's social status, and the literature [97] use features of the family place and the work place to find the interest of the mobile Internet users. These methods reveal a certain correlation between the social attributes of the user and the user's Internet behavior and can effectively analyze user behavior through social attributes. However, most of the data required by these methods belong to personal privacy and are hard to obtain.

Some people try to make use of data generated by users themselves, such as GPS information, browsing logs, search logs, etc., to analyze users' Internet habits and try to apply them to commercial push. Zhang et al. [97] display several characteristic behavior patterns that can guide service providers in application designing, operating, and marketing by the sequence of user web page access. Cole et al. [98] analyzed the sequence of user web page access and find that similar patterns of user activity are observed at both the cognitive and page use levels. Daniela uses network document data to design an unsupervised learning method for user portrait. These methods effectively depict users' online habits in content dimension. However, because these methods are only analyzed from the content dimension and ignore other relevant information in the data, such as time information, space information, and so on, the information of the dataset is not fully used.

So, many people begin to use context to analyze the user's Internet behavior. The literature [99] finds that the user's trajectory is associated with the APP usage pattern, and some geographical locations will affect the user's APP usage pattern. Jiang et al. use time information in data to predict the time of the MAS and microblog data and find behavior patterns. Han et al. [100] analyze user behavior by exploring the relationship between user interest and social characteristics. However, there are a few existing studies that combine the user's social attributes and the data generated by the user to analyze the user's Internet behavior habits.

This chapter uses the user's mobile phone brand data and the user's APP data to analyze the homogeneity and heterogeneity of users' Internet habits. In addition, this article is also an attempt at a joint analysis of the user's attributes and the data produced by the user itself.

In this chapter, we selected user detail records (UDR) data from a main mobile operator in Beijing, China, including 2,788,384 user entities. The total length of data time is more than 23 days, including the complete 3 weeks. It means that many types of Internet behavior will be recorded in UDR data and provides a guarantee for the richness and accuracy of the experimental results. Before data preprocessing, we made statistics on the proportion of mobile user brand. As shown in Table 4.7, it shows clearly that the top seven brands of mobile phones have accounted for more than 80% of the market share. Therefore, it is meaningful to analyze the relationship between APPs and these brands. On this basis, 3000 users are selected as samples for each brand concerned, which is used to quantitatively compare the behavior of brands of users.

We have selected some mobile users who have been using mobile phone services for over a year and believe that these users have more stable network usage behavior and habits, showing a more typical performance. We use the field structure of UDR data as shown in Table 4.8.

Table 4.7 The proportion of mobile brand

Mobile phone brand	Proportion
Apple	0.348431
Samsung	0.195126
MI	0.111555
Lenovo	0.050958
Huawei	0.046497
NOKIA	0.034173
Coolpad	0.02793
Others	0.18533

Table 4.8 Data structure before preprocessing

User_ID	Start_time	End_time	URL

Table 4.9 App information

APP name	Company	APP type
QQ	Tencent	Social
Baidu	Baidu	Search
Tieba	Baidu	Social
Zhidao	Baidu	Search
Taobao	Alibaba	Shopping
Alipay	Alibaba	Shopping
Tmall	Alibaba	Shopping
Amazon	Amazon	Shopping
Jingdong	Jingdong	Shopping
Meituan	Meituan	Shopping
Weibo	Sina	Portal
Sohu	Sohu	Portal
UC	UC	Portal
Youku	Youku	Video
IQiyi	IQiyi	Video
KuGou	KuGou	Music
Kuwo	Kuwo	Music
PPStream	IQiyi	Video
Letv	Letv	Video

Table 4.10 Data structure after preprocessing

User_ID	Start_time	APP

In the era of rapid development of smartphone Internet, the competition of APP is more intense. Therefore, users can choose different applications provided by different service providers for the same use of network behavior to achieve the same goal. Faced with tens of thousands of APP, we chose 20 different APP based on the market value of service providers and download number of App Store to simplify our experiments, as shown in Table 4.9. Through the URL information of UDRs, we can easily get user app usage record.

On this basis, we sorted the APP records of each user according to the start time of using app and then form single user discrete time series which shows the sequence of users' Internet behavior. According to discrete time series, user's transfer behavior between different APP is clear and we could know what another APP will be chosen after user leaving accessing APP. The process of recording from UDR to discrete time series is completed under Apache Spark. It is a distributed computing framework based on Hadoop and has obvious advantages in large-scale data processing. After data preprocessing, we use the field structure of the UDR data as shown in Table 4.10.

Fig. 4.33 Pictures of the
experimental frame, which
shows the relationship among
LSI, LMILP, and association
rule

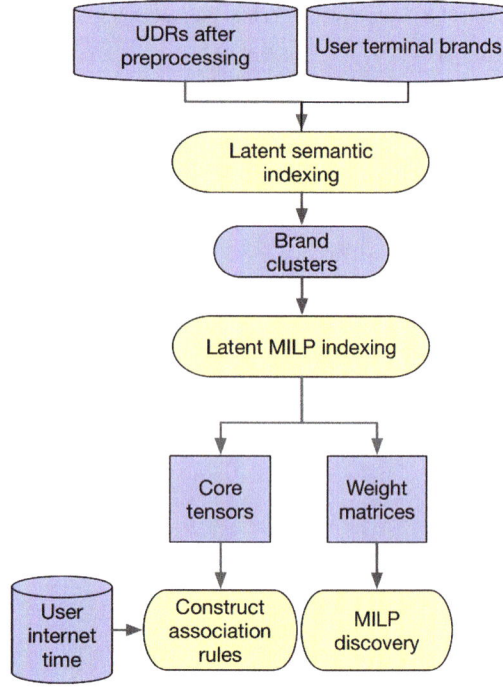

In this chapter, a correlation analysis scheme between mobile phone brand and APP is proposed. The program is divided into three parts. The first part mainly uses the LSI thematic model to find the correlation between brand and APP and the potential hidden theme. Meanwhile, we classify the brand of mobile phone based on the characteristics of mobile phone, which is called brand cluster. In the second part, we introduce the concept of MILP and propose a LMILPI model which aims to discover and extract the MILP hidden in the data. In this way, we can find MILP of different brand clusters. In the last part, we would like to find the change of MILP in different time periods so we try to analyze the existing phenomena by association rule mining to find the rules hidden in the dataset. The program is shown in Fig. 4.33.

To find the correlation between brand characteristics and APP features, many researchers have done a lot of work and achieved notable results. In this chapter, we try to analyze the correlation between brand characteristics and APP features from the perspective of topic models.

LSI, the abbreviation of latent semantic indexing, is one of the classic models in the theme model, which is often used to find the relationship between words through mass literature. When some words always appear in a document, there might be a relationship between them, in other words, these words can be regarded as semantically related. By LSI model, we can quickly analyze the homogeneity between words and get the relationship. At the same time, LSI can also find the

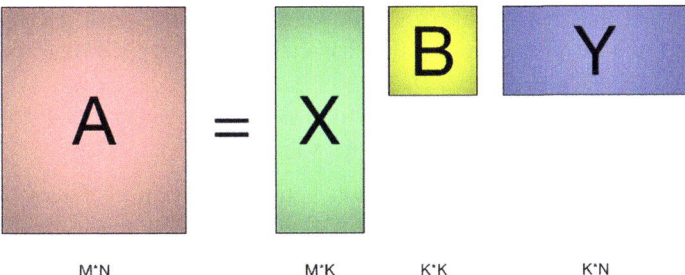

Fig. 4.34 A schematic diagram of the LSI model. *A* is a decomposition matrix. The left singular matrix *X* represents the correlation between the *M* word and the *K* semantic class. The right singular matrix *Y* represents the correlation between the *N* article and the *K* theme

homogeneity of the literature and the correlation between the literature and the vocabulary.

From the algorithm level, LSI is based on the singular value decomposition (SVD) method to get the subject of the text, as shown in Fig. 4.34.

In this article, we think that different kinds of APP are like words which have their own unique "semantics." And these unique "semantics" will match the strongly related "articles," that is, the mobile phone users of different brands. From the point of view of characteristic decomposition, the "semantic" is the feature of APP, and different APP has different characteristics. And different users also have different features. When user groups with different mobile phones are using the Internet, APP features attract the user group with matching feature. Therefore, we can use LSI to analyze the correlation between mobile phone brand and APP.

In our experiments, we randomly selected 30,000 users from UDRs to ensure the accuracy and effectiveness of the experiment. In each brand, we count the number of different APP visits by these users, and form statistica matrix X_{M*N}. M represents the number of brands, N represents the number of APP, and $X_M N$ represents the total usage count that users of the brand m use APP n. We processed the matrix X by logarithmic process and obtained the decomposition matrix A.

We use the SVD decomposition of the matrix A. Based on the size of the eigenvalues, we select the first three column eigenvectors to analyze.

This chapter mainly analyzes the relationship between the total usage count and the characteristic values and find the substitution and complementarity between different APP. Meanwhile, we divide the brand cluster and the APP cluster based on the hierarchical clustering method, and get the correlation between the brand and APP.

In the current analysis of group behavior, the researchers have found the following phenomena:

- Different users have different behaviors, that is, characteristics of all people are not exactly same.

- Some users' Internet behaviors are similar and they have same interest, that is, some users have the same or similar characteristics and typical behavior patterns [97].
- When users are on the Internet, the current using APP will have an impact on the APP which we will be using next time [101].

Based on the above phenomena, we can assume that the user's Internet behavior is determined by the user's own characteristics and the APP that is used now. So, we put forward the latent mobile Internet life personas indexing method based on the above conclusion, that is, LMILPI based on our assumption to find user's characteristics.

Through the LMILPI model, we can analyze the user portrait and find typical behavior patterns hidden in users' behavior. We call it mobile Internet life portrait (MILP) which could describe user's transfer behavior between different APP and do analysis on users. Next, we will elaborate on the modeling methods and algorithms of the model.

In modeling, based on Phenomenon 1, we think that user U has a different MILP of K, as P_k. P. And each MILP will have different degrees of influence on user's behavior according to different weights $\omega(P)$. The greater the weight, the stronger the impact of MILP on users. Based on Phenomenon 3, we believe that different MILP has different preferences for different APP, and this difference is affected by the current access of APP. MILP and the accessing APP, namely A_t, jointly decide the upcoming APP, namely A_{t+1}. We define the preference as the access intensity F, whose value is $f(P_k, A_t, A_{t+1})$ and the larger the value, the stronger the access to A_{t+1} preference under a priori conditions. Therefore, we put forward the following equation:

$$P(A_t, A_{t+1}) = \sum_k \omega(P_k) * f(P_k, A_t, A_{t+1}) \tag{4.68}$$

By Phenomenon 2, we can know that some user characteristics are the same or similar, which means that many users may have the same or similar MILP. Therefore, it is meaningful to analyze the MILP shared by users. To analyze the MILP shared by group users, we can transform formula 1 into tensors, as shown in formula 2, so we transform the MILP discovery problem of group users into a tensor decomposition problem.

$$P(U, A_t, A_{t+1}) = F(P, A_t, A_{t+1}) \times W(P, U) \tag{4.69}$$

Among them, $P(U, A_t, A_{t+1})$ is the statistical data, which is the tensor of the probability that the user U has access to A_t at the current moment and will access to A_{t+1} at the next time. $W(P, U)$ refers to the weight of K different MILPs for user U, and $f(P_k, A_t, A_{t+1})$ is the access intensity tensor made up of different MILPs.

At the algorithm level, to decompose the tensor, we combine the nonnegative tensor decomposition called NTF and the high-order singular value decomposition HOSVD method to deal with the data.

HOSVD, like SVD, is a singular value decomposition method for high-order tensors. It can decompose tensors into three factor matrices and a kernel tensor, and its formula is as follows:

$$\mathbb{R} = \chi \times_1 A \times_2 B \times_3 C \tag{4.70}$$

Combined with the 2 and 3 formula, we can calculate the tensor decomposition problem based on HOSVD algorithm, to fully guarantee the nonnegativity of tensors and the independence between features. The formula is as follows:

$$W(P, U) = AF(P, A_t, A_{t+1}0) = \chi \times_2 B \times_3 C \tag{4.71}$$

Based on the above modeling method and solution algorithm, we proposed the latent Mobile internet life indexing model to discover the potential MILP of the user.

First, we select 3000 users randomly for each brand cluster generated by LSI, and we can easily build user transition probability tensor P through their UDR data. Then we can decompose the probability tensor P of user transfer through the LMILPI model and get the weight of MILP $W(P, U)$ and access intensity tensor $F(P, A_t, A_{t+1})$. To expand access tensor, we can obtain the access K strength matrixes $f(A_t, A_{t+1})$ of different MILP.

In this chapter, we mainly compare the MILP of four different types of brand clusters and analyze the homogeneity and heterogeneity between MILP of different brand clusters.

Association rule mining is a rule-based machine learning algorithm, which can find interesting relationships in large databases. Its purpose is to use some metrics to identify strong rules that exist in a database. Association rules mining is used for knowledge discovery, not prediction. In this chapter, we apply the association rule algorithm to find out the rules between MILP extracted from different mobile brand users and users' main Internet time.

First, we analyze the user's online time. We divide 1 day into four periods, of which 0:00–6:00 are time period 1, 6:00–12:00 are time period 2, 12:00–18:00 are time period 3, and 18:00–24:00 are time period 4. We find out the most frequent time of the user's Internet access and mark it as the user's main Internet time T_u.

Then, we calculate the user weight matrix $W(P, U)$ obtained by the LMILPI model, and select the MILP with the largest weight as the main MILP of users, which is called P_u.

At the same time, we can get user's ID and brand information based on UDR data, so that we can get the UDR data of single user U into a set of vectors.

$$U \rightarrow (B_u, P_u, T_u) \tag{4.72}$$

Among them, B_u represents the user U_s mobile phone brand cluster, which can be obtained through the LSI model, and P_u refers to the main MILP of the user U, and T_u refers to the user's main Internet access time (Table 4.11).

Currently, the problem of seeking correlation among brand cluster, main Internet time, and main MILP becomes the problem of mining association rules for vectors. In the mining of association rules, it is very important to find a reasonable rule. In this chapter, we define the rules that need to be analyzed as follows:

$$(U, B_i, T_i) \Rightarrow (U, P_k) \tag{4.73}$$

In the mining of association rules, two indexes play a decisive role. They are support degree and confidence degree. The support degree represents the probability of the rule appearing in the whole dataset, and the confidence degree represents the conditional probability of the rule appearing in the whole dataset. In this article, we find relationships among brand cluster, main time, and main MILP mainly through support and confidence (Table 4.12).

We use the LSI theme model to analyze the relationship between the mobile phone brand and the APP. The singular values of the eigenvectors are obtained by the SVD algorithm as shown in Table 4.14. According to the singular value, we can find that the first three items of singular value account for most of the total value of

Table 4.11 SVD singular value results

Singular value
50.46841
3.164664
2.143987
0.853441
0.636626
0.505439
0.276828

Table 4.12 Clustering results of LSI theme model

Type	Mobile brand	APP information					
		Shopping	Portal	Video	Music	Social	Search
1	Mi	Taobao		IQiyi			
2	Huawei	Alipay	Weibo				
	Nokia	Meituan	Sohu	Youku	KuGou	QQ	Baidu
		Amazon	UC	Letv		Tieba	Zhidao
		Jindong					
3	Samsung						
	Lenovo	Tmall		PPStream	Kuwo		
	Coolpad						
4	Apple						

Table 4.13 The result of the left singular matrix analysis

APP name	Ratio	APP name	Ratio
QQ	−133.386	Weibo	−133.83
Baidu	−133.568	Sohu	−133.637
Tieba	−133.519	UC	−133.693
Zhidao	−133.251	Youku	−133.775
Taobao	−133.255	IQiyi	−132.844
Alipay	−133.665	KuGou	−133.598
Tmall	−132.774	Kuwo	−133.098
Amazon	−133.297	PPStream	−132.945
Jingdong	−133.402	Letv	−133.365
Meituan	−133.43		

Table 4.14 The result of the right singular matrix analysis

Mobile brand	Ratio	Mobile grand	Ratio
Apple	−216.15	Huawei	−218.119
Samsung	−217.682	Nokia	−217.848
Mi	−218.677	Coolpad	−218.241
Lenovo	−218.391		

the singular value. Therefore, we try to select the first three eigenvectors to analyze the relationship between the terminal brand and APP.

We found that the eigenvectors corresponding to the maximum singular value have a great relationship with the total amount of users' access. We count the total number of user's hits based on brands and compare it with the first column feature vector of the left singular vector. Surprisingly, the ratio between them is basically constant. In Tables 4.13 and 4.14, this result shows that the first column of eigenvalues indicates user's hits. We count the number of hits based on APP and compare the result with the first line of the right singular vector, and we get the same conclusion.

Then, we projected the two dimensions of the left singular vector and the right singular vector to the plane, as shown in Fig. 4.3. We can find that different mobile phones show different features. Several mobile phones are closer, indicating that they are more similar, and several mobile phones are far away, which means the gap between them is larger. APP also has the same phenomenon. To make a scientific analysis of it more specifically, we use hierarchical clustering algorithm to cluster it, and get several typical "brand clusters" and "APP clusters," as shown in Table 4.12.

Through LSI, the correlation of "brand cluster APP cluster" can be established, and they are roughly divided into the following parts:

• The potential relationship between mobile phone brands was found. LSI divides the cell phone into four main categories: "Mi," "HUAWEI & NOKIA," "Samsung & Coolpad & Lenovo," and "Apple." This result is very interesting because it is related to the market positioning of these mobile brands. In the 2015 smartphone sales share, Samsung, Apple, Huawei, and Mi account for the top four in the world mobile phone market (Market share is 24.8%, 17.5%, 8.4%, and 5.6%), and their products positioning and target customers are different. Mi is mainly aimed at users seeking cost-effective and young people. Apple is mainly for who like high quality products. HUAWEI are mainly used by business people who are always busy. Samsung have products for low income and middle income customers who have large free time. So, they are divided into four different clusters. Besides these main brands, "Lenovo" and "Coolpad" are learning the marketing model of "Samsung" and competing with it so the cluster called "Samsung & Coolpad & Lenovo" is build. This phenomenon also appears in the marketing strategy of HUAWEI and NOKIA, so the " HUAWEI & NOKIA " has been set up.

- From the internal relationship of APP cluster, the competition relationship between applications can be displayed. In the shopping class APP, the "Tmall," "Taobao," "Jingdong," and other shopping applications have a competitive relationship. They are usually not used by the user at the same time, which is different from the traditional impression that the purchase of goods requires more than one shopping software. From the results of LSI, each user has a customary shopping app, and mainly uses it online. This phenomenon is also reflected in the video class APP and music class APP. Because of the existence of exclusive copyright, a specific video product cannot be found in two different APP, and this also forms the competition pattern between them.

- There is a strong correlation between mobile phone brand and APP. For example, users of Mi prefer Taobao; users of Samsung & Coolpad & Lenovo" always use Tmall; and "HUAWEI & NOKIA" users use many different shopping APP. The reason for this phenomenon is the difference in user purchasing power. Users who use Mi mainly focus on cost performance, so they use Taobao to buy cost-effective products. Users of Samsung have begun to focus on the brand of goods, so Tmall, which has more brand stores, is more popular with them. Most of the users of Huawei are business people. They focus on shopping experience and aftersales service. So Jingdong and Amazon are chosen by them (Fig. 4.35).

In the video APP, IQiyi mainly plays what young people like to see, while Mi users are mainly young people, so Mi is related to IQiyi. PPS is the main player in TV play, so "Samsung & Coolpad & Lenovo" users who have free time, will use it. Youku and music have rich content, so business people "HUAWEI & NOKIA " prefer it.

These findings also provide mobile phone manufacturers with proposals for service recommendation and service pre-installation.

The recommendation strategy, which is more consistent with the needs of the user, can achieve a better marketing effect.

Based on LSI, we can make a reasonable classification of the user's mobile phone brand. To balance computation speed and experimental results, we select 3000 users from each brand cluster and use LMILPI models to find the homogeneity and heterogeneity of MILP among different brand clusters. To achieve this goal, we show the MILPs of different cluster in the form of a thermal diagram, as shown in Figs. 4.36, 4.37, 4.38, and 4.39.

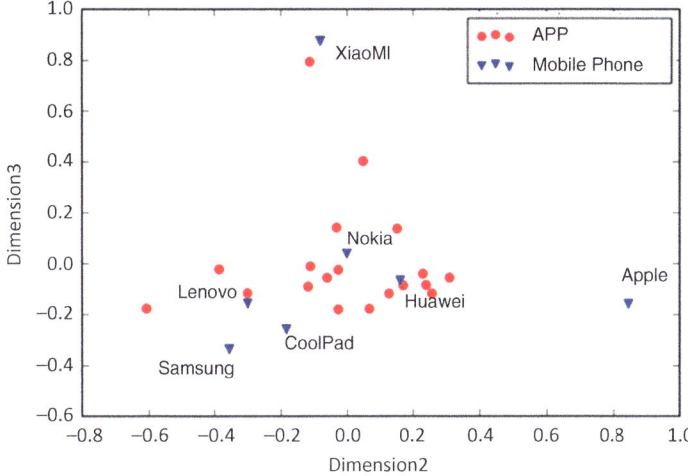

Fig. 4.35 We selected the singular vector of the second and the third dimension of the left and right singular matrices, and then projected them to the two-dimensional plane as shown in the figure

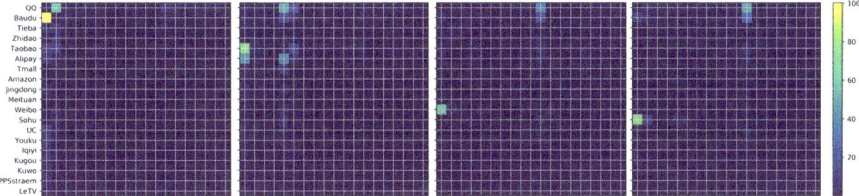

Fig. 4.36 MILPs of brand cluster 1

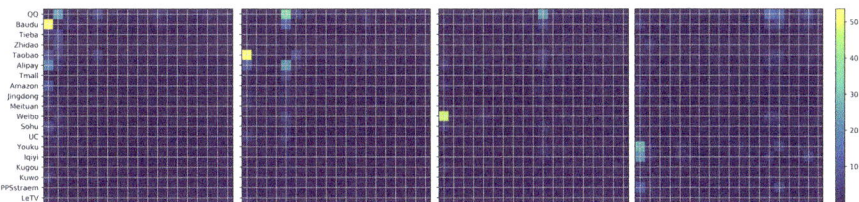

Fig. 4.37 MILPs of brand cluster 2

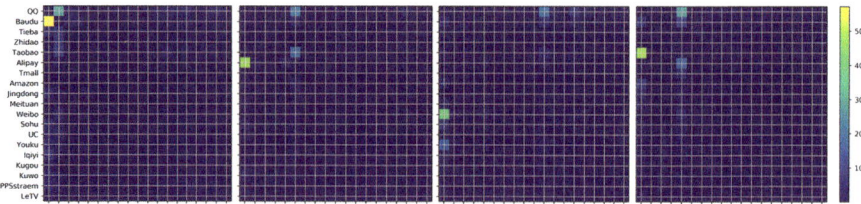

Fig. 4.38 MILPs of brand cluster 3

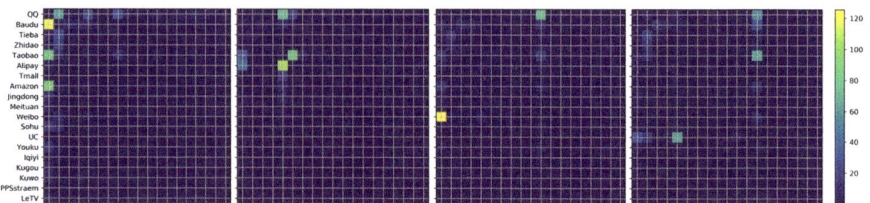

Fig. 4.39 MILPs of brand cluster 4, and Figs. 4.4, 4.5, 4.6, and 4.7 show the MLIPs extracted from the LMILPI model. We use thermodynamic diagrams to depict four typical MLIPs. In the diagram, each row represents the currently used app, A_t, and each column represents the app that will be accessed, called A_{t+1}, and its values show the tendency of the user from A_t to A_{t+1}. In the picture, the color tends to be yellow, indicating that this tendency is more intense. Tending to blue means a tendency to be weak

We have analyzed the results of the decomposition and found that there are similarities and differences between different types of MILP. First, we try to analyze the decomposition results of the first type of brand cluster users in Fig. 4.4.

The first MILP indicates that the user prefers to choose a "Baidu-QQ" cycle, that is, when users use QQ, they may visit Baidu next and then return to QQ, thus forming a cycle. Meanwhile, when users end up using other APP except Baidu and QQ, users are more likely to return to Baidu or QQ and re-enter this cycle. This phenomenon is consistent with the results obtained by the LSI model, that is, the two services of Baidu and QQ are divided into one category.

For the second MILP, the social APP such as QQ and the shopping APP, namely Taobao and Alipay, are more important than others. It is worth to note that, as you can see from the image, the user will have a large probability of using Alipay after using the Taobao. The two APP came from Alibaba. Taobao is an important shopping platform and Alipay is an electronic payment device. Users will use Alipay to pay money after they choose their favorite products in Taobao, which is also consistent with our reality. Meanwhile, besides the strong correlation between Taobao and Alipay, users still have the preference to switch from QQ to Taobao. This can also help Alibaba to develop advertising strategy, which is to push effectively with Tencent through QQ.

For the third MILP, like the first class of MILP, it forms a "Weibo-QQ" cycle. Weibo is a platform for everyone to express their views, similar to Twitter. For the fourth MILP, it indicates users who tend to use SOHU. This is a portal, carrying all kinds of contents and having access to all kinds of services.

According to four MILPs, we can find some interesting results. Baidu, Tencent, and Alibaba are the most important companies in China, and our results also show that. And Tencent, the owner of QQ, has the most abundant traffic.

When we have completed the internal analysis of the first type of brand cluster users, we try to analyze the similarities and differences between different brand clusters.

We find that mobile phone on Android has similar MILP. "Mi," "HUAWEI & NOKIA," and "Samsung & Coolpad & Lenovo" all have the Baidu-QQ cycle, the QQ-Taobao-Alipay cycle, and the Weibo-QQ cycle. However, there are differences among the fourth MILP. "Mi" tends to visit SOHU; "HUAWEI & NOKIA" tends to access APP classes such as Youku and IQiyi; and "Samsung & Coolpad & Lenovo" mainly visits Alipay.

However, there is a significant difference between Apple and the main MILP of these three types of brand clusters. The first MILP of Apple shows that users have a strong tendency to switch from Baidu, Taobao, Amazon to QQ. The second item has more emphasis on the connection between Taobao and Alipay than the first three. The third item is also Weibo-QQ. The last item emphasizes the importance of a browser called UC. It can be found that many Apple users use UC browser as an Internet access to access other APP.

Based on MILP, our analysis can give service providers reliable business advice. For example, it is appropriate for UC to cooperate with Apple and Tencent rather than Mi and Baidu. So, service providers can choose the right mobile phone manufacturers or other service providers to cooperate according to our results.

We calculate the support degree and confidence degree by the method of association rules and draw them as Fig. 4.40.

As you can see from pictures, MILP may have a change in different periods of time. For example, MILP1's support degree increases first and then decreases in brand cluster 1, and the confidence degree decreases. In addition, the comparison of the value of support degree between different MILP will produce different results. In cluster 2, the support degrees of MILP3 are less than MILP4 support in period 3, but in period 4, MILP3 is more than MILP4.

In combination with the phenomenon in LMILPI, we found that for cluster 1, cluster 2, and cluster 3, the first three MILPs are basically similar. However, from the point of association rule mining, we can find the difference between the three types of MILP.

First, for brand cluster 1, MILP1 and MILP4 can have a greater impact on users in four time periods. For brand cluster 2, MILP1 is in an important position, while for brand cluster 3, MILP2 plays an important role. This phenomenon shows that even though the MILP decomposed by LMILPI model is similar, the importance of different MILP in different cluster is different. This is the complement of LMILPI algorithm.

The analysis of confidence degree is helpful for service providers to make different recommend strategy to brand clusters in different time periods. For

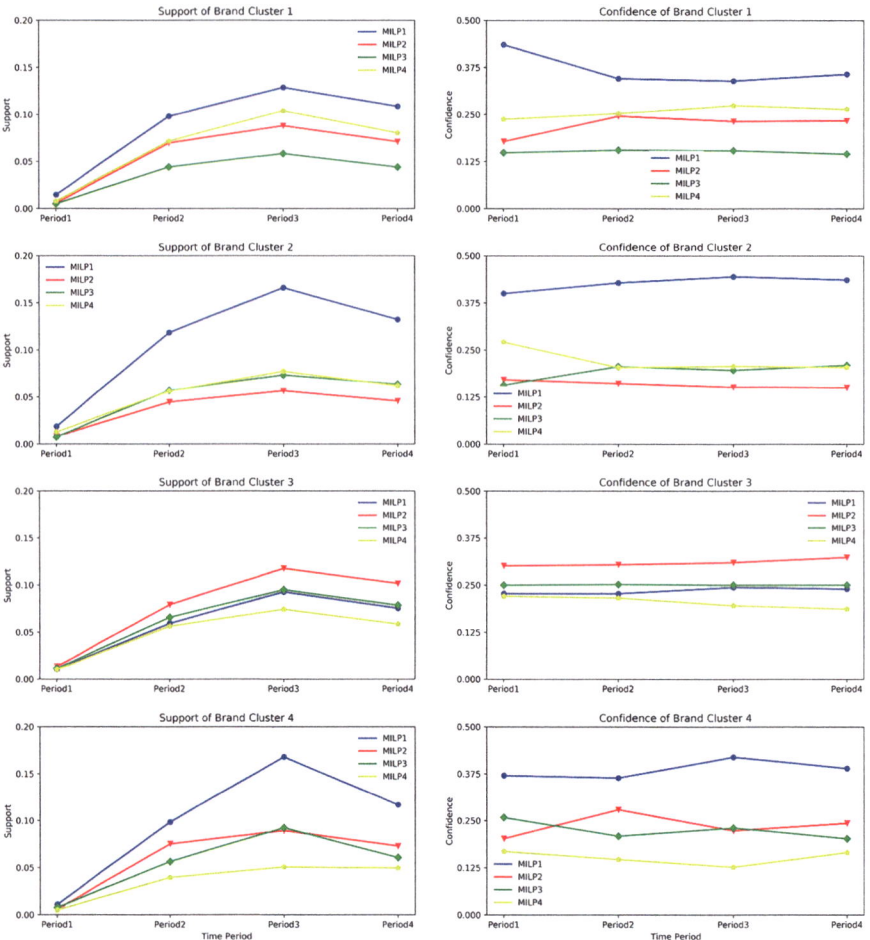

Fig. 4.40 Four images on the left show the support of different MILP in the brand cluster 1 to the brand cluster 4. The four graphs on the right show the confidence of different MILP in the brand cluster 1 to the brand cluster 4

example, for brand cluster 4, in period 1 and period 3, the proportion of MILP1 and MILP3 is larger, so we should push Baidu, Taobao, Amazon, and Weibo information when users use QQ. While in period 2, when users use QQ, pushing Baidu, Taobao, and Alipay information is suitable and it is the time that Alipay and Taobao are used at the same moment which means people want to do online shopping at period 2. Based on these results, users in clusters can get the appropriate APP according to the time information and the user's main MPLI.

Based on support degree, we can analyze the MILP weight matrix of the user and make up for the LMILPI model we proposed. With confidence degree, we can comprehensively analyze the relationship among mobile brand, user's Internet time, and user's main MILP to provide service providers appropriate recommendation strategies based on time.

Based on LSI, LMILPI, and association rules mining, we propose a framework called user portrait analysis framework to explore typical user behavior. The framework mainly studies the law of user transfer between different APPs and analyzes user behavior based on mobile phone brand.

In our framework, LSI will guide the cooperation between service providers and mobile phone manufacturers. LMILPI mainly guides the cooperation between service providers. Association rule mining will help service providers to know which period is the most appropriate time to recommend. For example, based on LSI, UC can get the information that Apple is the best cooperative business in kinds of mobile manufacturers and UC will known Weibo and Sohu are its competitor. Based on LMILPI, many people will use UC to do online shopping, so UC should have close relationship with Taobao. By the way, Alibaba, the company of Taobao, has bought UC in recent years. And based on association rules, provide services of UC knowns UC should be pushed to users who use iPhone when they hit Taobao and afternoon is the most suitable time to recommend. Therefore, service providers can formulate appropriate recommendation strategies through our framework and maximize the effectiveness of recommendations and mobile manufacturers can improve user experience by preloading APP.

The comprehensive development of mobile Internet brings huge business opportunities and profits to mobile phone manufacturers and service providers. However, the expansion of market size is also accompanied by competition. Reasonably finding and analyzing the theme of mobile Internet life using different mobile phone brands can help mobile producers and service providers to seize and increase market share. In this chapter, we propose an analysis framework based on the hidden theme model to mining the theme features of mobile Internet users.

The main contributions of this article include the following:

- Through the classic theme model LSI, the relationship between different brand mobile phones and different APP is found.
- The LMILPI model is proposed and the mobile Internet life personas of users with different mobile phone brands is found and compared based on the LMILPI model.
- The relationship between the mobile phone brand, the user's main Internet time, and the user's main MILP is analyzed.

The above work can help mobile phone manufacturers and service providers understand the user's Internet behavior patterns at a certain level. However, in real life, although most users access the Internet through mobile operators, there are still some users who can access the Internet through the wireless network, such as WIFI, which cannot be obtained. Therefore, it cannot fully reflect the portrait of the user.

In addition, this chapter uses the HOSVD algorithm in the calculation of the LMILPI model. Although the algorithm provides stable decomposition results, the consumption of computing resources is relatively large, so it is difficult to carry out large-scale data calculation. It is our next step to improve the HOSVD algorithm and reduce its consumption of resources.

In this chapter, we extract and compare the users' MILP from different brand phones based on feature analysis and topic discovery and do user portrait analysis based on this. This may be ideal. The theme of Internet life is a broad and inaccurate concept. It covers many aspects of human behavior in the network. What we do at present is only a small part of it.

Of course, our current work also provides some ideas for social computing. Other sociological factors, such as user occupation and socioeconomic status, will have an impact on user behavior patterns. Despite the limitations, we believe this research can provide a framework for a new mobile Internet service model.

References

1. Pepper, R.: Cisco visual networking index (VNI) update global mobile data traffic forecast update. Mobile Word Congress (2013)
2. Neto, J.L.D., Yu, S., Macedo, D.F., Nogueira, J.M.S., Langar, R., Secci S.: ULOOF: a user level online offloading framework for mobile edge computing. IEEE Trans. Mob. Comput. **17**(11), 2660–2674 (2018)
3. Tang, L., He, S.: Multi-user computation offloading in mobile edge computing: a behavioral perspective. IEEE Netw. **32**(1), 48–53 (2018)
4. Ketyk, I., Kecsks, L., Nemes, C., Farkas, L.: Multi-user computation offloading as multiple knapsack problem for 5G mobile edge computing. In: European Conference on Networks and Communications, pp. 225–229 (2016)
5. Chen, X., Jiao, L., Li, W., Fu, X.: Efficient multi-user computation offloading for mobile-edge cloud computing. IEEE/IEEE/ACM Trans. Netw. **24**(5), 2795–2808 (2016)
6. Sankaran, C.B.: Data offloading techniques in 3GPP Rel-10 networks: a tutorial. IEEE Commun. Mag. **50**(6), 46–53 (2012)
7. Zhuo, X., Gao, W., Hua, S., Hua, S.: An incentive framework for cellular traffic offloading. IEEE Trans. Mob. Comput. **13**(3), 541–555 (2014)
8. Lee, K., Lee, J., Yi, Y., Rhee, I., Chong, S.: Mobile data offloading: how much can WiFi deliver? In: International Conference, pp. 1–12 (2010)
9. Balasubramanian, A., Mahajan, R., Venkataramani, A.: Augmenting mobile 3G using WiFi. In: International Conference on Mobile Systems, Applications, and Services. pp. 209–222 (2010)
10. Han, B., Hui, P., Kumar, V.S.A., Marathe, M.V., Shao, J., Srinivasan, A.: Mobile data offloading through opportunistic communications and social participation. IEEE Trans. Mob. Comput. **11**(5), 821–834 (2012)
11. Izumikawa, H., Katto, J.: Rocnet: spatial mobile data offload with user-behavior prediction through delay tolerant networks. In: Wireless Communications and Networking Conference, pp. 2196–2201 (2013)
12. Hui, P., Crowcroft, J., Yoneki, E.: Bubble Rap: social-based forwarding in delay tolerant networks. IEEE Trans. Mob. Comput. **10**(11), 1576–1589 (2011)
13. Han, B., Hui, P., Kumar, V.S.A., Marathe, M.V., Pei, G., Srinivasan, A.: Cellular traffic offloading through opportunistic communications: a case study. In ACM Workshop on Challenged Networks, pp. 31C38 (2010)
14. Li, Y., Su, G., Hui, P., Jin, D., Su, L., Zeng, L.: Multiple mobile data offloading through delay tolerant networks. IEEE Trans. Mob. Comput. **13**(7), 1579–1596 (2014)

15. Andreev, S., Pyattaev, A., Johnsson, K., Galinina, O.: Cellular traffic offloading onto network-assisted device-to-device connections. IEEE Commun. Mag. **52**(4), 20–31 (2014)
16. Andrews, J.G., Claussen, H., Dohler, M., Rangan, S., Reed, M.C.: Femtocells: Past, present, and future. IEEE J. Sel. Areas Commun. **30**(3), 497–508 (2012)
17. Ramaswamy, V., Das, D.: Multi-carrier macrocell femtocell deployment-a reverse link capacity analysis. In: Vehicular Technology Conference Fall, pp.1–6 (2009)
18. Golaup, A., Mustapha, M., Patanapongpibul, L.B.: Femtocell access control strategy in UMTS and LTE. IEEE Commun. Mag. **47**(9), 117–123 (2009)
19. Elsawy, H., Hossain, E., Camorlinga, S.: Traffic offloading techniques in two-tier femtocell networks. In: IEEE International Conference on Communications, pp. 6086–6090 (2013)
20. Mukherjee, A., De, D.: Low power offloading strategy for femtocloud mobile network. Int. J. Eng. Sci. Technol. **19**(1), 260–270 (2016)
21. Paris, S., Martisnon, F., Filippini, I., Lin, C.: A bandwidth trading marketplace for mobile data offloading. In: IEEE INFOCOM, pp. 430–434 (2013)
22. Zhuo, X., Gao, W., Cao, G., Dai, Y.: Win-coupon: an incentive framework for 3G traffic offloading. In: IEEE International Conference on Network Protocols, pp. 206–215 (2011)
23. Dong, W., Rallapalli, S., Jana, R., Qiu, L.: ideal: incentivized dynamic cellular offloading via auctions. In: IEEE INFOCOM, pp.755–763 (2013)
24. Bulut, E., Szymanski, B.K.: WiFi access point deployment for efficient mobile data offloading. ACM SIGMOBILE Mob. Comput. Commun. Rev. **17**(1), 71–78 (2013)
25. Shenker, S.: Fundamental design issues for the future internet. IEEE J. Sel. Areas Commun. **13**(7), 1176–1188 (2002)
26. Duan, L., Huang, J., Shou, B.: Duopoly competition in dynamic spectrum leasing and pricing. IEEE Trans. Mob. Comput. **11**(11), 1706–1719 (2012)
27. Lv, L.: Link prediction on complex networks. JESTC **39**(5), 651–661 (2010)
28. Li, Q., Zheng, Y., Xie, X., Chen, Y., Liu, W., Ma, W.Y.: Mining user similarity based on location history. In: ACM SIGSPATIAL International Conference on Advances in Geographic Information Systems, p. 34 (2008)
29. Lv, M., Chen, L., Chen, G.: Mining user similarity based on routine activities. Inf. Sci. **236**(1), 17–32 (2013)
30. Lv, L., Zhou, T.: Link prediction in weighted networks: the role of weak ties. Europhys Lett. **89**(1), 18001 (2010)
31. Feng, M., Mao, S., Jiang, T.: Base station on-off switching in 5g wireless networks: approaches and challenges. IEEE Wirel. Commun. **24**(4), 46–54 (2017)
32. Budzisz, L., Ganji, F., Rizzo, G., Marsan, M.A.: Dynamic resource provisioning for energy efficiency in wireless access networks: a survey and an outlook. IEEE Commun. Surv. Tutorials **16**(4), 2259–2285 (2014)
33. Zhao, T., Liu, Q., Chen, C.W.: QoE in video transmission: a user experience-driven strategy. IEEE Commun. Surv. Tutorials **19**(1), 285–302 (2017)
34. Gomez, G., de Torres, E., Lorca, J., Garcia, R., Perez, Q., Arias, E.: Assessment of multimedia services QOS/QOE over LTE networks. In: International Conference on E-Business and Telecommunications, pp. 257–272 (2012)
35. Khan, M.A., Toseef, U.: User utility function as quality of experience (QOE). In: Proceedings of the ICN, pp. 99–104 (2011)
36. Wu, J., Zhang, Y., Zukerman, M., Yung, K.N.: Energy-efficient base-stations sleepmode techniques in green cellular networks: a survey. IEEE Commun. Surv. Tutorials **17**(2), 803–826 (2015)
37. Hu, R.Q., Qian, Y.: Macro-Femto Heterogeneous Network Deployment and Management. Wiley, Hoboken (2013)
38. Li, W., Zheng, W., Xie, Y., Wen, X.: Clustering based power saving algorithm for self-organized sleep mode in femtocell networks. In: International Symposium on Wireless Personal Multimedia Communications, pp. 379–383 (2012)
39. Han, T., Ansari, N.: Enabling mobile traffic offloading via energy spectrum trading. IEEE Trans. Wirel. Commun. **13**(6), 3317–3328 (2014)

40. Zhu, Y., Kang, T., Zhang, T., Zeng, Z.: QoS-aware user association based on cell zooming for energy efficiency in cellular networks. In: IEEE International Symposium on Personal, Indoor and Mobile Radio Communications, pp. 6–10 (2014)
41. Oh, E., Son, K., Krishnamachari, B.: Dynamic base station switching-on/off strategies for green cellular networks. IEEE Trans. Wirel. Commun. **12**(5), 2126–2136 (2013)
42. Jiang, H., Yi, S., Wu, L., Leung, H., Wang, Y., Zhou, X., Chen, Y., Yang, L.: Data-driven cell zooming for large-scale mobile networks. IEEE Trans. Netw. Serv. Manag. **15**(1), 156–168 (2018)
43. Son, K., Kim, H., Yi, Y., Krishnamachari, B.: Base station operation and user association mechanisms for energy-delay tradeoffs in green cellular networks. IEEE J. Sel. Areas Commun. **29**(8), 1525–1536 (2011)
44. Shaikh, J., Fiedler, M., Collange, D.: Quality of experience from user and network perspectives. Ann. Telecommun. **65**(1–2), 47–57 (2010)
45. Fiedler, M., Hossfeld, T., Tran-Gia, P.: A generic quantitative relationship between quality of experience and quality of service. IEEE Netw. **24**(2), 36–41 (2010)
46. Ding, M., Wang, P., Lpez-Prez, D., Mao, G., Lin, Z.: Performance impact of LoS and NLoS transmissions in dense cellular networks. IEEE Trans. Wirel. Commun. **15**(3), 2365–2380 (2016)
47. Quesada, I., Grossmann, I.E.: An LP/NLP based branch and bound algorithm for convex MINLP optimization problems. Comput. Chem. Eng. **16**(10–11), 937–947 (1991)
48. Liu, B., Zhao, M., Zhou, W., Zhu, J.: Flow-level-delay constraint small cell sleeping with macro base station cooperation for energy saving in HetNet. In: Vehicular Technology Conference, pp.1–5 (2015)
49. Zhu, Y., Kang, T., Zhang, T., Zeng, Z.: QOS-aware user association based on cell zooming for energy efficiency in cellular networks. In: IEEE International Symposium on Personal, Indoor and Mobile Radio Communications, pp. 6–10 (2014)
50. Barabasi, A.L.: The origin of bursts and heavy tails in human dynamics. Nature. **435**(7039), 207 (2005)
51. Becker, R., et al.: Human mobility characterization from cellular network data. Commun. ACM **56**(1), 74–82 (2013)
52. Liu, Y., et al.: Uncovering patterns of inter-urban trip and spatial interaction from social media check-in data. PloS One. **9**(1), e86026 (2014)
53. Becker, R., et al.: Human mobility characterization from cellular network data. Commun. ACM **56**(1), 74–82 (2013)
54. Peng, C., et al.: Collective human mobility pattern from taxi trips in urban area. PloS One. **7**(4), e34487 (2012)
55. Guo, D.: Visual analytics of spatial interaction patterns for pandemic decision support. Int. J. Geogr. Inf. Sci. **21**(8), 859–877 (2007)
56. Belik, V., Geisel, T., Brockmann, D.: Natural human mobility patterns and spatial spread of infectious diseases. Phys. Rev. X. **1**(1), 3103–3106 (2001)
57. Liang, X., et al.: Unraveling the origin of exponential law in intra-urban human mobility. Sci. Report. **3**(10), 65–65 (2013).
58. Kerner, B.S.: Introduction to modern traffic flow theory and control: the long road to three-phase traffic theory. Phys. Today **63**(3), 53 (2010)
59. Candia, J., et al.: Uncovering individual and collective human dynamics from mobile phone records. J. Phys. A Math. Theor. **41**(22), 224015 (2008)
60. Longini, I.M., et al.: Containing pandemic influenza at the source. Science **309**(5737), 1083–1087 (2005)
61. Isaacman, S., et al.: Human mobility modeling at metropolitan scales. In: Davies, N. (ed.) Proceedings of the 10th International Conference on Mobile Systems, Applications, and Services, June, Windermere, p.239C252. ACM, New York (2012)
62. Gonzlez, M.C., Hidalgo, C.A., Barabsi, A.-L.: Understanding individual human mobility patterns. Nature **453**(7196), 779–782 (2008)

63. Cheng, Z., et al.: Exploring millions of footprints in location sharing services. In: Nicolov, N., Shanahan, J.G. (eds.) International Conference on Weblogs and Social Media. Barcelona, Catalonia. AAAI, Menlo Park (2011)
64. Rhee, I., et al.: On the levy-walk nature of human mobility. IEEE/ACM Trans. Netw. **19**(3), 630–643 (2011)
65. Liang, X., et al.: The scaling of human mobility by taxis is exponential. Physica A Stat. Mech. Appl. **391**(5), 2135–2144 (2012)
66. Liu, Y., et al.: Understanding intra-urban trip patterns from taxi trajectory data. J. Geogr. Syst. **14**(4), 1–21 (2012)
67. Li, Q., et al.: Mining user similarity based on location history. In: Samet, H., Shahabi, C., Wolfson, O. (eds.) ACM SIGSPATIAL International Symposium on Advances in Geographic Information Systems, p.34. ACM New York (2008)
68. Xiao, X., et al.: Finding similar users using category-based location history. In: Agrawal, D., Zhang, P. (eds.) ACM SIGSPATIAL International Symposium on Advances in Geographic Information Systems, pp. 442–445. ACM, New York (2010)
69. Ying, J.C., et al.: Mining user similarity from semantic trajectories. In: Zhou, X., Lee, W.C. (eds.) ACM SIGSPATIAL International Workshop on Location Based Social Networks, p. 19–26 (2010)
70. Lv, M., Chen, L., Chen, G.: Mining user similarity based on routine activities. Inf. Sci. **236**(1), 17–32 (2013)
71. Yuan, J., Zheng, Y., Xie, X.: Discovering regions of different functions in a city using human mobility and POIs. In: Yang, Q. (ed.) Proceedings of the 18th ACM SIGKDD International Conference on Knowledge Discovery and Data Mining, Beijing, pp. 186–194. ACM, New York (2012)
72. Pei, T., et al.: A new insight into land use classification based on aggregated mobile phone data. Int. J. Geogr. Inf. Sci. **28**(9), 1988–2007 (2014)
73. Jiang, S., et al.: Clustering daily patterns of human activities in the city. Data Min. Knowl. Disc. **25**(3), 478–510 (2012)
74. Guo, D.: Flow mapping and multivariate visualization of large spatial interaction data. IEEE Trans. Vis. Comput. Graph. **15**, 6 (2009)
75. Kang, C., et al.: Exploring human movements in Singapore: a comparative analysis based on mobile phone and taxicab usages. In: Koonin, S.E., Wolfson, O.E. (eds.) Proceedings of the 2nd ACM SIGKDD International Workshop on Urban Computing, Chicago, pp.1–8. ACM, New York (2013)
76. Sui, Z.W., Wu, L., Liu, Y.: Study on interactive network among Chinese cities based on the check-in dataset. Geogr. Geo-Inform. Sci. **29**(6), 1–5 (2013)
77. Wang, F., Antipova, A., Porta, S.: Street centrality and land use intensity in Baton Rouge, Louisiana. J. Transp. Geogr. **19**(2), 285–293 (2011)
78. Barabasi, A.L.: The origin of bursts and heavy tails in human dynamics. Nature **435**(7039), 207 (2005)
79. Brockmann, D., Hufnagel, L., Geisel, T.: The scaling laws of human travel. Nature **439**(7075), 462–465 (2006)
80. Song, C., et al.: Modelling the scaling properties of human mobility. Nat. Phys. **6**(10), 818–823 (2010)
81. Zhou, C., et al.: TCB: a feature transformation method based central behavior for user interest prediction on mobile big data. Int. J. Distrib. Sens. Netw. **12**(10), 1550147716671256 (2016)
82. De Montis, A., Caschili, S., Chessa, A.: Commuter networks and community detection: a method for planning sub regional areas. Eur. Phys. J. Spec.Top. **215**(1), 75–91 (2013)
83. Gao, S., et al.: Discovering spatial interaction communities from mobile phone data. Trans. GIS **17**(3), 463–481 (2013)
84. Yuan, Y., Raubal, M., Liu, Y.: Correlating mobile phone usage and travel behavior case study of Harbin, China. Comput. Environ. Urban. Syst. **36**(2), 118–130 (2012)
85. Wu, W., et al.: Oscillation resolution for mobile phone cellular tower data to enable mobility modelling. In: Zaslavsky, A., Chrysanthis, P.K., Becker, C. (eds.) IEEE, International

Conference on Mobile Data Management, July. Brisbane, pp. 321–328. IEEE Computer Society, Washington, D.C., 2014

86. Wang, W., et al.: A comparative analysis of intra-city human mobility by taxi. Physica A Stat. Mech. Appl. **420**, 134–147 (2015)

87. De Montis, A., et al.: The structure of inter-urban traffic: a weighted network analysis. Environ. Plann. B. Plann. Des. **34**(5), 905–924 (2007)

88. Newman, M.E.: The structure of scientific collaboration networks. Proc. Natl. Acad. Sci. **98**(2), 404–409 (2001)

89. Wu, L., et al.: Incorporating human movement behavior into the analysis of spatially distributed infrastructure. PloS One. **11**(1), e0147216 (2016)

90. Delvenne, J.C., Yaliraki, S.N., Barahona, M.: Stability of graph communities across time scales. Proc. Natl. Acad. Sci. **107**(29), 12755–12760 (2010)

91. Schaub, M.T., et al.: Markov dynamics as a zooming lens for multiscale community detection: non clique-like communities and the field-of-view limit. PloS One. **7**(2), e32210 (2012)

92. Pfitzner, D., Leibbrandt, R., Powers, D.: Characterization and evaluation of similarity measures for pairs of clusterings. Knowl. Inf. Syst. **19**(3), 361–394 (2009)

93. Schieber, T.A., et al.: Quantification of network structural dissimilarities. Nat. Commun. **8**, 13928 (2017)

94. Mcdaid, A.F., Greene, D., Hurley, N.: Normalized mutual information to evaluate over-lapping community finding algorithms. Comput. Sci. arXiv preprint arXiv:1110.2515. 2011

95. Sarwar, B., et al.: Application of dimensionality reduction in recommender system-a case study. Technical report, Minnesota Univ Minneapolis Dept of Computer Science, Minneapolis, 2000

96. Luo, S., Morone, F., Sarraute, C., Travizano, M., Makse, H.A.: Inferring personal economic status from social network location. Nat. Commun. **8**, 15227 (2017)

97. Zhang, X., Wang, C., Li, Z., Zhu, J., Shi, W., Wang, Q.: Exploring the sequential usage patterns of mobile Internet services based on Markov models. Electron. Commer. Res. Appl. **17**, 1–11 (2016)

98. Cole, M.J., Hendahewa, C., Belkin, N.J., Shah, C.: User activity patterns during information search. ACM Trans. Inf. Syst. **33**(1), 1 (2015)

99. Trestian, I., Ranjan, S., Kuzmanovic, A., Nucci, A.: Measuring serendipity: connecting people, locations and interests in a mobile 3G network. In Proceedings of the 9th ACM SIGCOMM Conference on Internet Measurement, pp. 267–279. ACM, New York, 2009

100. Han, X., Wang, L., Crespi, N., Park, S., Cuevas, A.: Alike people, alike interests? Inferring interest similarity in online social networks. Decis. Support. Syst. **69**, 92–106 (2015)

101. Zhao, Z.D., Yang, Z., Zhang, Z., Zhou, T., Huang, Z.G., Lai, Y.C.: Emergence of scaling in human-interest dynamics. Sci. Rep. **3**(12), 3472 (2013)

Chapter 5
Mobile Data Application in Smart City

Abstract This chapter focuses on mobile networks: The term smart city can be defined as an environment that uses currently available technologies to improve living conditions by offering access to information about parameters that enhance lives of its inhabitants. These parameters can range from status of their education and employment possibilities, available utilities for citizens, transportation information, energy consumption information, health-related issues, water and air quality monitoring, waste management, and other relevant information that potentially benefit the community. Smart city is regarded as the future direction of urban development. Some applications of smart city are as follows: (1) Make public service to become more intelligent. (2) Make urban traffic to become more convenient. (3) Ensure Internet security. The key technologies of smart city include city-level smart operation framework, city-wide sensing infrastructure, M2M telecommunication, and city open data. These technologies will promote the development and evolution of smart city. Therefore, mining mobile data and understanding the underlying mobile Internet lifestyle are applications of smart city and also provide direction for the development of smart city.

Keywords Smart city · Parameters · Urban development · Internet security · Smart operation framework · M2M

5.1 Lifestyles Mining for Urban City Areas

5.1.1 Introduction

China has become the largest market for mobile communications in the world [1]. (As of June 2017, people spend an average of 26.5 h a week online, of which 96.3% of these users are online via mobile phones [2].) Online data is gradually replacing the traditional social surveys and interview to understand the living habits of individuals in society. At the mobile Internet technology level, the ever-increasing global coverage of mobile cellular networks and huge data generation have provided support for the prosperity of the mobile Internet. In the combined effect of the above

© Springer Nature Switzerland AG 2019   179
H. Jiang et al., *Mobile Data Mining and Applications*, Information Fusion and Data Science, https://doi.org/10.1007/978-3-030-16503-1_5

factors, although the PC Internet has become increasingly saturated, the mobile Internet has been tremendously developed. The coverage and usage of large-scale mobile network provide a new motivation for recording the online life and social activities of urban people. Hence, understanding the underlying mobile Internet lifestyle is crucial to understand the personal behavior and social dynamics of the big data era. This is also an effective way for city planners to better understand the dynamics of modern cities, including cultural boundaries, Internet security, and even economic conditions.

Due to cultural, social-economic conditions, and other constraints, different regions of the population lifestyle obviously there are differences. Current research is based primarily on telephone interviews or survey data, which are always accompanied by relatively large time and financial costs. In addition, due to the limited time and space dimensions, these data do not accurately cover the range of human Internet activities.

The traditional notion of popularity for understanding the lifestyles of individuals in cyber-society is that people in metropolitan cities are active on the Internet, preferring social networks and shopping; in small cities, people are quiet and limited in Internet life. In order to validate a wide range of urban understanding, in this work, we consider the space in which the users are located and the Internet content they visit. We propose a solution to this "user-geo-Internet content" problem by extracting the typical characteristics of users considering high-dimensional mobile Internet contextual access content.

We think that people's Internet lifestyle is related to the Internet content they visit. In our current work, the style of individuals accessing content in the Internet is considered as a weighted combination of multiple qualitative lifestyles. The method uses the UDRs of the mobile Internet to infer people's network lifestyles. Since the data usage behavior is very common in daily life and is spontaneously generated by users accessing base stations (BSs), it accurately reflects the human space location information and network access behavior. After constructing the contextual information tensors for users to access mobile Internet content, we use the improved high-order singular value decomposition (HOSVD) method to find out the sequence features of these potential access contents. The extracted features provide a definition of the lifestyle of the user's network associated with a particular lifestyle, and they typically represent typical access characteristics of a particular group of people, such as game entertainment players. We chose Beijing as a representative of a big city in mainland China and Jinhua as a representative of a small city. We found out the Internet lifestyle in different sized cities and regions. In addition, similar to urban spatial structure, in combination with the spatial and geographical factors, we extracted the reliable spatial structure of network cities in both places. This method provides effective guidance for extracting complex sequence patterns in high-dimensional space.

The rest of Sect. 5.1 is structured as follows: In Sect. 5.1.2, we review related study work on social computing and mobile Internet mining and list some literature. In Sect. 5.1.3, introduce the framework including data preprocessing and model building, contextual information building and discovery of behavioral features.

In Sect. 5.1.4, describes and analysis experimental results including discovery of behavioral characteristics and regional Internet lifestyles.

5.1.2   Related Work

The result of researchers' research has greatly enriched the study of human behavior dynamics. Anonymous call detail records (CDR) is one of the main methods in using data to capture urban dynamics. In Milan's case, Parwez et al. analyzed abnormal behavior of mobile wireless network, and they compared the detected anomalies with the real information to verify their correctness [3]. Pulselli et al. show the possibility of using data to study urban activities and they use images to show the intensity of urban activity and evolution [4]. Reades et al. studied the dramatic changes in the activity of locals and tourists on weekdays and weekends in six different locations in Rome and proposed algorithms for clustering similar geographical locations [5]. Calabrese et al. speculate on the origins of people attending a special event in Boston [6]. The literature examines behavioral differences between tourists and locals in New York [7]. Because service providers regularly collect service business, planning, and billing policies, the resources needed to analyze these data are small. However, CDRs have some limitations. First, they are sparse in time because records are generated only when conversations are made, and second, CDRs will not record the data of users who are failed to make conversations, making the source of regional survey data incomplete.

In the mobile data mining method for mobile phones, we focus more on the extraction of typical features from network users, as shown in the following sections.

In the traditional method of feature mining, the research focuses more on the extraction and analysis of single-level features of users. Luo et al. designed a spherical frame by combining the structure hole features of social network models and proved that there is a strong correlation between the characteristics of human network life patterns and social status [8]. At the same time, the user's online behavior has also been shown to have an important connection with individual consumer performance [9]. They have proved the strong connection between user network behavior and the consumption of the real society and market behavior. Some studies consider the use of models to discover user characteristics. Park et al. proposed an eigen model, indicating that the eigenvectors of the user's transfer matrix can provide detailed information for the user's mobile mode [10]. Cole et al. designed a method based on Markov chain to discover the different behavior patterns by using individual sequence for webpage access [11]. It can be used to distinguish and represent different tasks. Zhang et al. used hidden Markov model and multi-state model to infer the usage patterns of mobile Internet users' apps [12]. Based on the characteristics of individual mobility, location, and travel behavior, Qiao et al. proposed a prediction framework for studying the behavior of individuals and groups of people [13]. Kawazu et al. used hidden Markov model to analyze various potential states of users for Internet surfing [14]. The advantages and disadvantages of the above methods are related to the statistical learning model used. The method of matrix and tensor decomposition starts from the individual behaviors and according to the founding theme of the composition of online behavior, Shafiq et al. proposed user trust and correlation matrix, and through this extraction method they found that the relevant information in the user network has strong correlation with the relevant information in the social network [15]. Jiang et al. proposed a new framework of tensor decomposition, which predicted the multi-directional behavior of time and the behavior patterns in MAS and Weibo data [16]. Wu et al. used the statistical characteristics of user network access as a basis for optimizing network deployment and found that the radius of user activation in space obeys a lognormal distribution [17]. In the meantime, the literature proposes a framework for using family and work-related contexts to identify user interests, and enriches traditional user behavior by introducing user interest [18]. However, to the best of our knowledge, this article will, for the first time, discuss the different economic regions from the perspective of individual behavior in the network under the conditions of regional differentiation between the typical network user online context mode features.

5.1.3 Study Framework

This section introduces the overall research process of users' mobile Internet behavior patterns which is shown in Fig. 5.1, including the following three steps: the first step is data preprocessing, mainly to organize user access time series sorted by the session time, the series consist of user ID, Internet session time, and user access APP types. Then, we construct the transfer probability matrix based on user access content. Fitting through multi-state models, the content time sequence forms a context-based transfer distribution. Finally, we can extract the core user characteristics and weights through the improved HOSVD, and get the composition of user mobile Internet lifestyle in cities of different scale through mobile Internet user pattern discovery.

5.1.3.1 Data Preprocessing and Model Building

In this study, we use UDRs from mobile cellular networks that contain fields of user ID, content information, session time, etc. All user IDs are protected by encryption. The dataset was collected from Beijing and Jinhua, China, including 2,788,384 and 1,519,661 individual users, respectively, and it spanned 23 days in 2015. Based on the dataset, we randomly selected 7000 users from the two cities for quantitative research. In order to ensure the sample data to be more reliable and representative, we screen out mobile users who have used mobile phones for more than a year for assuming that behaviors of these users have more stability and representation. Field structure of our UDRs is shown in Table 5.1.

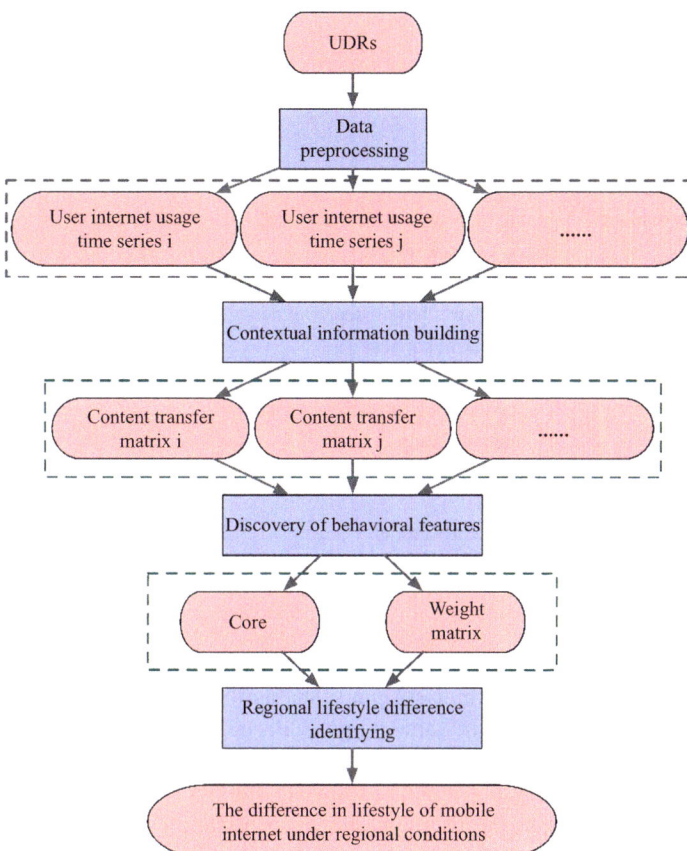

Fig. 5.1 Study framework

Table 5.1 Data structure
before preprocessing

Fields	Description
uid	An encrypted telephone number uniquely indicating a mobile user
stime	The time that a user begins to access the BS for data usage
URL	Uniform resource locator names that a mobile user accesses at the corresponding start time

We think a normal mobile user can only visit one target application during a conversation period. To make sure no duplicate information was included in the sampled users, we filtered out records that did not coincide with the start and end times of the user's conversations, ensuring that users only accessed one application or site at a time and then sampled.

Table 5.2 Data structure before preprocessing

Services	Description
Social	Instant communication, video voice service, broadcast online communication service
Shopping	Online shopping service
Portal & news	Portal and news access
Downloads & clouds	Application store downloads or P2P downloading services
Search	Internet search service
Travel	Travel and tourism resources acquisition
Entertainment	Games, live entertainment, online video
Forum	Online forum services
Life	Service to meet and assist the needs of life, weather, cooking guide
Mail	Mailbox service for mobile devices
Ad	Mobile phone ads
Fiction	Online reading

Table 5.3 Data structure before preprocessing

User ID	Start time	APP label

In the era of rapid growth of the smartphones, competition in the service market is even more fierce. Therefore, users can select applications provided by different service providers to achieve the same purpose. Based on Apple Inc.'s classification of mobile applications in App Store, we divide the services extracted from URL into 12 groups and give each group a tag, as shown in Table 5.2.

Based on this, we sort out users' accessing information and construct the time series of users' accessing application types. Structure of each record in the series is shown in Table 5.3.

The above process is implemented by Apache Spark, which is a distributed computing framework based on Hadoop, and has obvious advantages in large-scale data processing [19].

5.1.3.2   Contextual Information Building

In order to construct context information of user's online behavior, this chapter uses a multi-state model to obtain the transfer matrix of user behavior. This model is a tool of a continuous time stochastic process that allows individuals to move between a limited number of discrete states. Assume that time t is an individual state. The state corresponds to the category that the individual belongs to $S(t) \in \{1, 2, \ldots, N\}$, and the discretized interclass transfer strength can be defined as follows:

$$q_{rs}(t, z(t)) = \lim_{\Delta t \to 0} P(S(t + \delta t)) = (s|S(t))/\Delta t \tag{5.1}$$

By calculating the time series of edge transfer, the transfer matrix Q of APP categories can be obtained, which is used to describe the transformation intensity between the two groups. In line r of the matrix, the element in column s is, which can be used to represent the state transfer intensity of an individual transferred from APP r to APP s. The state transfer intensity represents the frequency of the transition from the APP r to the APP s in a single time series. According to the state transfer strength matrix Q, the state transition probability matrix P can be fitted to describe the individual switching between several states. The fitting method for P matrix, the row r column s elements can get this answer: when $r = s$, represents the transition probability from APP r to APP s. The state transfer probability matrix P obtained from the context information construction will serve as a further research basis and provide input.

We have applied the msm package in R that we already have in CRAN as a tool to train a polymorphic model, which was published by Jackson. In the training content setting, exact time is set to TRUE, which also means that the start time can be assumed to represent the exact time course conversion. obstype is set to 2, which means that the conversion time of the process is the exact conversion time, the user keeps the previous observation at a certain period of time until the current observation, and does not allow the change of state in the middle of the observation interval, which is in line with our expected.

From the training results of the multi-state model, we obtain the state transition distribution of each user. As mentioned above, the probability of state transition is more convincing and meaningful than the original transition intensity. Given the information on the average length of stay and transition probability matrix used in the training of polymorphic models, we chose the probability matrix as a basis for further research and provided input, given that we are more concerned with user information about the relationships among the services.

5.1.3.3 Discovery of Behavioral Features

For the discovery of potential characteristics of user mobile Internet context behavior, we believe that human activity can be represented by three-dimensional tensor Y, namely "user-transition matrix" structure, N is set to the number of mobile Internet APP tags, for target users, there are

$$Y \cong G \times_1 A^{(1)} \times_2 A^{(2)} \times_3 A^{(3)} = \sum_{p=1}^{P}\sum_{q=1}^{Q}\sum_{r=1}^{R} g_{pqr} A_p^{(1)} A_q^{(2)} A_r^{(3)} \qquad (5.2)$$

At the same time,

$$G \times_1 A^{(1)} \times_2 A^{(2)} \times_3 A^{(3)} = [G; A^{(1)}, A^{(2)}, A^{(3)}] \qquad (5.3)$$

G is the transfer probability matrix of the N kinds of APP, with k of them, which describe the potential characteristics of the individual's potential mobile Internet lifestyle. In order to balance the interpretability and accuracy of the results, we set the K to 4 in the experiment. $A^{(n)}$ represents the decomposition result of tensor along the N direction. Considering the objective function, we can get the following:

$$\|Y - [G; A^{(1)}, A^{(2)}, A^{(3)}]\| = \|\text{vec}\,(Y) - \left(A^{(3)} \otimes A^{(2)} \otimes A^{(1)}\right)\text{vec}\,(G)\| \quad (5.4)$$

Thus, G should subject to:

$$G = Y \times_1 A^{(1)T} \times_2 A^{(2)T} \times_3 A^{(3)T} \quad (5.5)$$

So the square of the objective function is converted to:

$$
\begin{aligned}
&\|Y - [G; A^{(1)}, A^{(2)}, A^{(3)}]\|^2 \\
&= \|Y\|^2 - 2\langle Y, [G; A^{(1)}, A^{(2)}, A^{(3)}]\rangle + \|[G; A^{(1)}, A^{(2)}, A^{(3)}]\|^2 \\
&= \|Y\|^2 - 2\langle Y \times_1 A^{(1)T} \times_2 A^{(2)T} \times_3 A^{(3)T}, G\rangle + \|G\|^2 \\
&= \|Y\|^2 - 2\langle G, G\rangle + \|G\|^2 \\
&= \|Y\|^2 - \|Y \times_1 A^{(1)T} \times_2 A^{(2)T} \times_3 A^{(3)T}\|^2 \quad (5.6)
\end{aligned}
$$

So the problem is transformed:

$$\max \|Y - [G; A^{(1)}, A^{(2)}, A^{(3)}]\| \Leftrightarrow \min \|Y \times_1 A^{(1)T} \times_2 A^{(2)T} \times_3 A^{(3)T}\|^2 \quad (5.7)$$

By using the idea of alternating solution, the problem can be transformed into a subproblem:

$$\max \|A^{(n)T} W\| \quad (5.8)$$

It is subject to $W = Y_{(n)}\left(A^{(3)} \otimes A^{(2)} \otimes A^{(1)}\right)$. To solve the optimization problem of the tensor decomposition, we adopt high-order orthogonal iteration (HOOI). This method is first proposed by Lathauwer [20], and the specific performance of HOOI-Tucker decomposition is provided by TensorLy toolkit. The aggregation of kernel tensors along the three different mode decomposition results, HOSVD will become a nonnegative matrix decomposition derived form, that is:

$$Y \cong G \times \omega \quad (5.9)$$

ω is a k-dimensional coefficient vector that represents the user's preference for each lifestyle and records the amount of the individual under a single typical feature. In order to reveal and compare the lifestyles usually followed in different cities, we first set the vector of activities of the residents to the single matrix of each city. We define:

$$Y = [p_1, p_2, \ldots, p_n]^T, n \in \{1, 2, \ldots, N\} \tag{5.10}$$

Among them, p_n is the set of transfer probability of APP n, in form, $p_n = [p_{1n}, p_{2n}, \ldots, p_{nn}]$. In the decomposed results, G represents the contextual characteristics of APP access behavior. Their total number is k, which together constitutes the user's mobile Internet lifestyle. The weight represents the occupancy of the different characteristics of the individual under different regional conditions, that is, the user's more inclined way of mobile Internet life.

5.1.4 Results Analysis

This section focuses on making a detailed explanation of the mobile Internet lifestyles (MILS) for regional users from the results of the tensor decomposition method.

5.1.4.1 Discovery of Behavioral Characteristics

In the analysis of large-scale data, the individual behavior is considered to be formed through the combination of user characteristics. Therefore, we try to use the tensor decomposition to study the individual characteristics of users and analyze the user's MILS.

In this chapter, improved HOSVD is used to extract the feature of user behavior, so as to construct a complete and accurate user network features. In the experiment, we got the typical user characteristics that differ among Beijing and Jinhua, as shown in Figs. 5.2 and 5.3.

First, we decompose the tensor of Jinhua users. By decomposing the above tensors, we find four kinds of basic user preference behavior features. In Fig. 5.3 the value of users' transfer behavior is depicted by different colors, while a darker

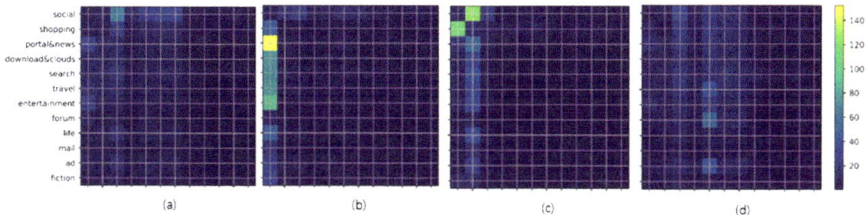

Fig. 5.2 Context characteristics of users accessing mobile Internet in Beijing

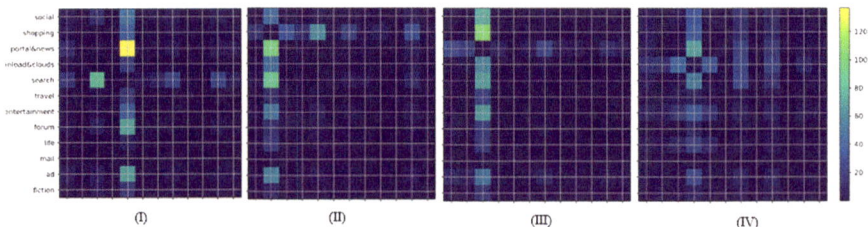

Fig. 5.3 Context characteristics of users accessing mobile Internet in Jinhua

yellow means a higher value, and a darker blue means a lower value. We find the below four features:

The first feature indicates that users prefer to use search-type content. From the figure, we can see that users prefer to flow from portal news, forums, and advertisement types to search categories, and they also tend to link to portal news content. At the same time, the picture also shows that users have the "portal news–search" cycle of this feature, that is, users will switch back and forth between portal news and search while accessing other types of mobile Internet content during the handover process.

The second feature shows the characteristics of users' access behavior of shopping content, in which users link to shopping-class mobile Internet content from portals, search, entertainment, and advertising. In addition, users also tend to link to search category content.

The third feature shows the characteristics of users' access behavior of portal news content, that is, users tend to access portal news content from shopping, downloading, searching, entertainment, and other content.

The fourth feature indicates other characteristics of users, e.g., the user has a small amount of access to mobile Internet content such as downloading, entertainment, living, and advertisement. In the above four types of Internet content, the downloading service is used as the main traffic port.

At the same time, we also analyze the results of Beijing. We found that Beijing users have notable features in the switching of Internet content. That is, users have a strong dependence on social content. We decompose the user's transfer probability tensor, and get four typical features of user's access.

The first type of user features represents the user of whom the interest of the network content is low, and the most common behavior is accessing from social class to portal news class.

The second type of user profile indicates the users' access behavior of shopping-class content, that is, the user tends to shift from other content to the shopping class, and then the user switches back and forth between the social and shopping-class content.

The third type describes the scenario in which a user accesses social content. In which users prefer to migrate content that is mainly portrayal and entertainment to social content. However, these users have low tendency to roll out.

The fourth feature shows that users have similar preferences among all Internet content such as news portal, search, travel, and other Internet services. This phenomenon shows that there is also a preference for moving from search, forums, and advertising to search content among these users.

Within the small deviation allowed, we can make a definition of MILS for the core tensors of the two cities in the above figure. For the four elements mentioned in Jinhua:

MILS 1 Searcher, the inflow and outflow of search class app is much larger than other types among these users, so these individuals are keen to use a variety of search software to understand the world.

MILS 2 Shopaholic, while users show special preference for shopping apps.

MILS 3 Current affairs observers, who are passionate about current events and have a high interest in news, probably correspond to middle-aged and elderly people in the region.

MILS 4 Casual player, they are less dependent on the network, occasionally use the download, entertainment for traffic expenses.

However, in Beijing, the results show the following:

MILS 1 Busy bee, these users have low overall access intensity, only for occasional news and search content, which is similar to the behavior of white-collar workers.

MILS 2 Gregarious bird, these users are the spokespersons of social networks. From the intensity of the columns, it is clear that social apps are their entrance to all kinds of content. At the same time, their online life cannot be separated from social applications such as WeChat or Weibo.

MILS 3 Sociable shopper, the access behavior of the shopping and social software of these users forms a circle, and they are more inclined to switch between these two types of software.

MILS 4 Business man, they are not interested in the Internet usage. They only search on the Internet to understand other content, such as forums, travel, etc.

We can get some meaningful comparisons between the two nuclear tensors. As a developing city, in Jinhua, the composition of the user's Internet access is relatively simple and it is difficult to form an obvious circle. However, due to the specific online behavior of users, a combination of various potential features can be found. Based on the results of feature discovery, service providers can formulate special recommendations in the region, for example, placing a product recommendation in a news application or providing attractive game links under a search engine, etc. This will result in better traffic. In Beijing, as a well-developed city, users' behavior is composed of mixed usage of multiple APPs, obvious circle structures such as social-shopping loops can be found. The presence of busy bee and business man makes it even more challenging for service owners to seize market opportunities. As a result, we found that users in Beijing are more concentrated when surfing the

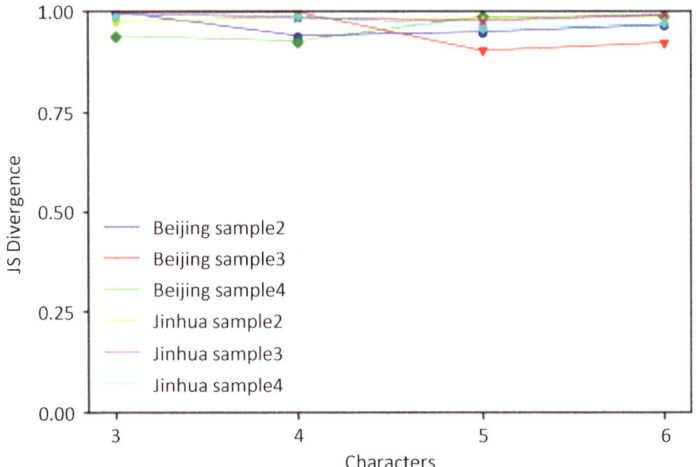

Fig. 5.4 Comparison of sampling similarity in two cities based on JS divergence

Internet. Therefore, we recommend that regulators supervise social applications as the main carrier so as to reduce the expenditure of other applications.

Taking into account the accuracy of the results, we reselect the sample data from the current data, and apply the mentioned feature extraction method to the MILS discovery. The resampling is still selected randomly. The number of samples is the same as that of the previous experiment. The results of Figs. 5.2 and 5.3 are based on JS similarity of divergence. The method was proposed by Shannon and later published in the literature [21]. Its result is symmetrical and the distance is between 0 and 1, while a larger value means a larger similarity. The similarity among sampling results of the two cities is shown in Fig. 5.4. As it can be seen in Fig. 5.4, the decomposition results of different samples are roughly similar when the number of MILS is 3–6, which also statistically shows the composition of the mobile Internet surfing features that exist in the two regions.

5.1.4.2 Regional Internet Lifestyles

Based on the weighted results of the two cities, we need to divide the distribution of MILS weights more granularly into urban areas. Analyzing the spatial pattern of human activities can provide a high understanding of the patterns of daily activities. And also analyzing human's initiative may give rise to an awareness of the patterns that improve their daily activities. In cellular networks, users always visit the nearest base station, so that the location data accuracy is about 300–500 m [22]. Therefore, each user's location can be approximated by the sequence of locations of the BSs. It is interesting to explore the geographical distribution of people with a small range

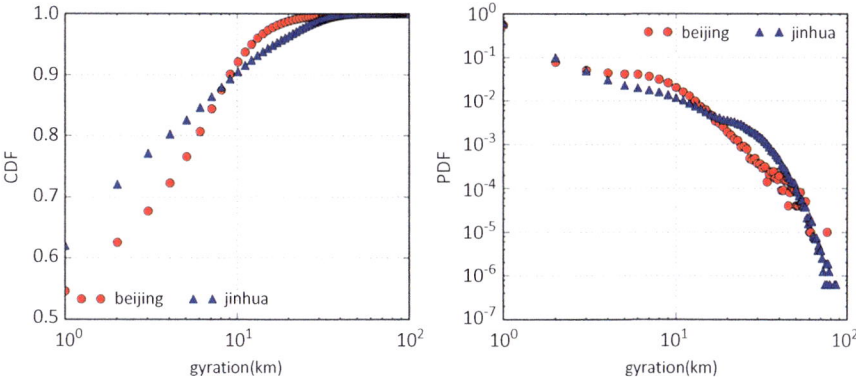

Fig. 5.5 Statistics of the circle radius of the users in two cities. Note the cumulative density function diagram of the active radius on the left side of the user and the probability density function diagram on the right on behalf of the user

of daily activities (R) [23]. Considering the mobility of users, we study the radius of users' activity. The activity radius of user u in space r is defined as follows:

$$r_g^u = \sqrt{\frac{1}{n_c^u} \sum_{i=1} \left(r_i^u - r_{cm}^u\right)^2} \tag{5.11}$$

Wherein, represents the centroid position of the movable range of u during the observation period. Figure 5.5 shows the cumulative density function (CDF) and probability density function (PDF) of mobile user gyration in both regions. We can see that mobile users have limited spatial activities region in both Beijing and Jinhua. In particular, about 55% of users in Beijing have a gyration of less than 1 km, with a ratio of 62% in Jinhua and 90% in Beijing and Jinhua with less than 10 km of gyration. In the double logarithmic axis, the PDF of the gyration of users in Beijing and Jinhua shows two kinds of distribution, as their characteristics are different. In the distribution characteristics of Beijing, users are 10 km away from each other. For the user groups whose gyration radius is more than 10 km, we find that the probability density shows a straight line. For the users whose gyration radius is less than 10 km, the probability density shows an exponential curve. In Jinhua, the distance among users is 12 km, and the areas on both sides of the boundary show a linear distribution, but the slope has significant difference.

Due to the local policies, local customs, and cultural influences, cities have natural cultural and economic differences [24]. The fine-grained urban individual users' online lifestyle can use the administrative area as an entry point. It can be seen from the users' gyration radius statistics in both places that more than 90% of the mobile user's gyration radius is within 10 km, while the average administrative area of Jinhua is 1215.8, and 1025.7 in Beijing, which can cover most users' activities radius, so a hypothesis about the region can be set, that is, the majority

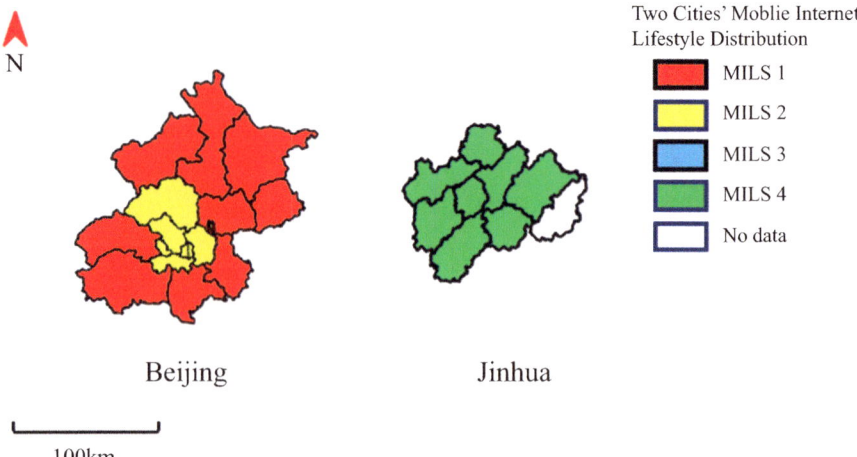

Fig. 5.6 Two of the most representative MILS maps under the administrative division of the city. Note that the map on the left represents the city of Beijing, and the map on the right represents Jinhua

of mobile users will not move out of their own administrative area. According to this hypothesis, we found the distribution characteristics of the weights of MILS in the region. If we use the largest proportion in the region as the representative of the entire administrative region, as shown in Fig. 5.6, it is obvious that the most important proportion of Jinhua is MILS 3, corresponding to "casual player," which means that in the entire city, all regions of the mobile Internet individuals are mainly reflected in such a network personality. However, different results were found in Beijing. In Beijing, the distribution of the main weights seems to be polarized. For example, in Changping District, the main user group is "busy bee," which infers that the overall strength of the Internet access is not very high. In the five urban districts and Changping District, users showed more enthusiasm for social networking and higher network activity. These two results are consistent with the conclusion of Vincente and Lopez [25] that they are in the European Union-27. In the country survey, users in the urban areas or in the more developed regions were found to have a higher degree of Internet access. However, we found that in relatively underdeveloped areas, users of the Internet are not representative of the characteristics. Figure 5.6 shows the MILS representative statistics of the two places.

Specifically, the results of the ratio of MILS types in the two regions are shown in Figs. 5.7 and 5.8.

In Fig. 5.8, a specific distribution of MILS can be obtained. GDP per capita and population density are important factors for the measurement of regional development. Therefore, we have obtained the per capita GDP and population density of Beijing in 2015 [26]. The 16 districts are divided into two categories, one includes Dongcheng, Xicheng, Chaoyang, Haidian, Fengtai, Shijingshan, Shunyi,

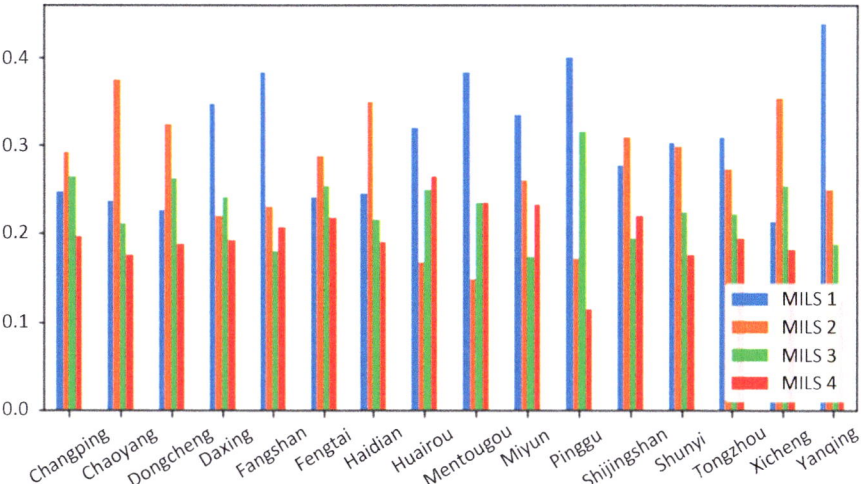

Fig. 5.7 Intra-regional occupation ratio in Beijing's MILS

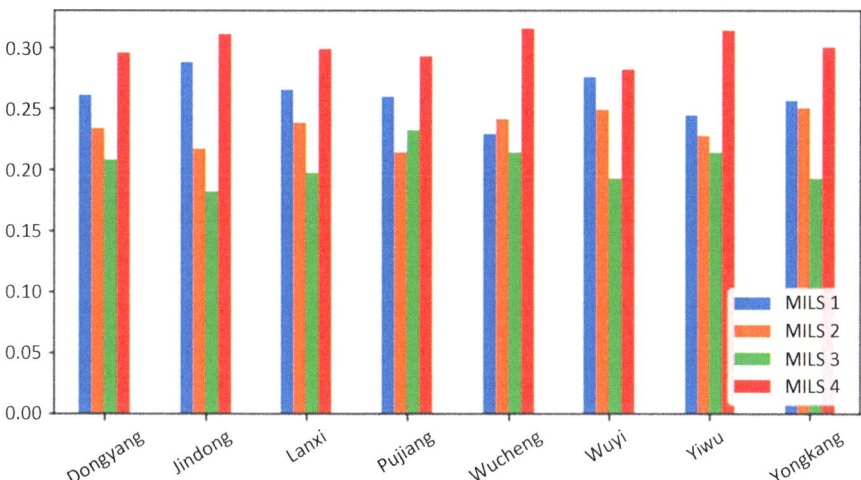

Fig. 5.8 Intra-regional occupation ratio in Jinhua's MILS

and Daxing, represented as the economically developed regions, and the other one includes Yanqing, Miyun, Pinggu, Mentougou, Fangshan, Huairou, Tongzhou, and Changping, represented as the less developed areas. As it can be seen from the distribution diagram, the top five regions of MILS 1 are Yanqing, Pinggu, Mentougou, Fangshan, and Daxing, while the areas with the highest number of MILS 2 are Chaoyang, Xicheng, Haidian, Dongcheng, and Shijingshan, so we can conclude that users in MILS 1 (who are not enthusiastic about Internet life) mostly live in more developed areas, and users in MILS 2 (who are keen on surfing the

Internet and love social issues) mostly live in underdeveloped areas. This further validates that the social preference is more likely to be affluent, which is in line with the conclusion reached by Luo et al. [8] that the socio-economic status of people with a wide social circle is high.

There are two interesting areas in the image, namely Daxing and Changping District. Only in these two regions, the population characteristics are opposite to the statistics. However, it can be found from the social behavior that there are a larger number of non-resident residents than the other regions (1,963,000 and 1,562,000, respectively). And their network character is related to the potential for development in these two regions (both of which have been incorporated into the Beijing Urban Development Area).

On the other hand, Jinhua has become the largest commodity distribution center in the world due to the existence of Yiwu District (with a GDP of 104.6 billion yuan in 2015 [27]). In recent years, the development of e-commerce has provided a new source for the development of small commodities, thus leading to a high status of MILS 2 (Shopaholic) in Jinhua city. However, as a developing city (GDP of 340.65 billion yuan in 2015 compared to 2.301559 trillion yuan in Beijing), different urban development processes (developed metropolitan and developing commercial cities) and urban economic development process led to the formation of a major pattern of MILS distribution in Jinhua (MILS4 > MILS1 > MILS2 > MILS3). Information APPs (search, current affairs), message carriers (shopping APPs and product launching websites), and entertainment are the major lifestyles for self-employed individuals in the small-commodity market. The "Yiwu mode" is widely used by people in Internet life, a manifestation of the influence of highly developed regions in ordinary cities. One more effort needs to be attempted, namely to find the influence significant representation in the city, that is, the "Yiwu Mode" in many developing regions, which portrays the city's unique demographic, and provide important guide for mobile operators, APP service providers, and government policy design.

The distribution of derivative weights is a hallmark of the everyday activity patterns of individual networks, enabling us to systematically compare the behavior characteristics of human network users in different geographic regions. The results of the analysis reveal that the impact of human activity space in both cities is worth discussing. First of all, a considerable part of the network life in these two cities is very limited. This unique mode of activity may reflect a number of social issues, which is mainly related to low-income migrant communities.

For most people in both cities, a small scale of network activity is usually enough to meet daily Internet needs, which is in line with the government and operators' goals because small-scale access can make network regulation and market policies more accurate.

However, under the conditions of economic development and the differences between North and South, the typical characteristics of the network users in two cities and their geographical distribution are shaped differently. This context-based universal feature not only provides an exciting market channel for operators and service providers, but also a network regulatory understanding of the ability to guide the app for the government. In the context of the weak ties and structural holes in social networks [8], it is important to discover the network dominance of mobile applications.

5.2   Behavior-Aware Collaborative Caching in Mobile Networks for Smart City Applications

5.2.1   Introduction

The term smart city in [28] refers to the use of information and communication technologies to sense, analyze, and integrate key information from core systems in operating cities. As the development of smart city, the load of wireless data transmission will increase rapidly [28, 29], and cellular networks in cities are under great pressure due to the exponential growth of wireless data traffic. A recent report by Cisco projects that mobile data traffic will reach 49.0 exabytes per month by 2021 [30]. This traffic growth will lead to greater pressure on the transmission link. Meanwhile, mobile networks, which can be regarded as a promising technology that provides computation and storage resources at remote radio heads (RRHs), have become a trend for the solution of exponentially increasing mobile data traffic.

Most prior research uses "homogeneous" caching methods, in which different RRHs either cache the same popular content or cache with the same probabilistic placement policy [31–34]. However, the data traffic and user access patterns at different RRHs exhibit spatial diversity, which is ignored by the "homogeneous" caching methods. The state-of-the-art RRH caching methods can be classified into collaborative caching methods and non-collaborative caching methods. In non-collaborative caching, RRHs with a certain amount of cache contents respond to the requests of users individually [35, 36]. When an RRH is requested for content not in its local cache, the RRH asks the Internet for a copy. The characteristics of the RRHs and users, namely the location of the RRHs and the content requested by the users, are considered in the collaborative caching method. Thus, RRHs with independent caching placement serve users collaboratively, which means that, when request is received, a certain number of RRHs can serve this request collaboratively in different ways. In [37, 38], the authors use the locations of the RRHs to determine the collaborative relationship, while in [39], RRHs that cache the same content are able to provide service collaboratively. However, most of the mentioned works ignore the original access behavior, which is depicted by users' multiple features, namely content and location features.

In this chapter, we propose a GUCC method for fifth-generation (5G) cellular networks. The main idea of the GUCC method is the utilization of group user behavior in the design of the collaborative relationship and the caching placement.

The main contributions of this chapter are summarized as follows:

- We design a collaborative relationship model based on group user behavior. We use a content and location similarity network to depict the group behavior. The

WSNF algorithm is implemented to solve the fusion problem of the large-scale similarity network, and user groups are formed based on the fused network. The collaborative relationship among the RRHs is established according to the user groups they serve.

- We propose a caching placement scheme that simultaneously maximizes the hit rate and minimizes the transmission cost. An optimization problem is solved according to MCMC theory to obtain the caching placement. A refreshing policy with low overhead is introduced for practical application.
- We apply the method to a large dataset generated from real networks and present numerical results regarding the hit rate and transmission cost performance.

The rest of Sect. 5.2 is structured as follows: In Sect. 5.2.2, we analyze a UDR dataset and observe several characteristics. Section 5.2.3 presents the architecture of the caching system. Section 5.2.4 explains the GUCC method. In Sect. 5.2.5, we present numerical results to evaluate the performance of the proposed method. Finally, Sect. 5.2.6 extends the discussion and concludes the chapter.

5.2.2   Group-Based User Behavior Analysis

Human behavior has been long analyzed [40–42], and the long-tail distribution and power-law distribution of user behavior have been verified from different perspectives, namely visited locations and visit times. In this section, we attempt to further explore access behavior.

We conduct the analysis with a dataset generated in Jinhua, China. The data span 23 days and mainly include detailed user network access information.

First, we obtain an overview of the diversity of users with respect to the content and location features. The diversity of users is defined as the number of unique contents or locations they have visited. As shown in Fig. 5.9, users' access behavior has a limited range of location and content features, similar to numerous previous works focused on the analysis of human behavior [43]. Few users show extreme content or location diversity, while most users access few contents and locations.

Despite the skew of the popularity, different people may prefer different contents. To illustrate the preference diversity among users, we calculate the Kullback–Leibler divergence (KLD) [44] of different users. We use the UDRs of the 4000 users who generated the most records, and calculate the KLD of every two users according to the content access behavior. The distribution of the KLD is shown in Fig. 5.10. Most of the divergence values among users are distributed around 8, which is a relatively high value in the distribution. Thus, despite the fact that all the users are visiting limited locations and contents, the preferences of users are different.

Now, we focus on the predictability of users' behavior. We use the Shannon entropy and the conditional entropy to individually evaluate the distribution of the content and location features. A lower entropy indicates lower uncertainty, and a decrease in entropy represents an improvement in predictability.

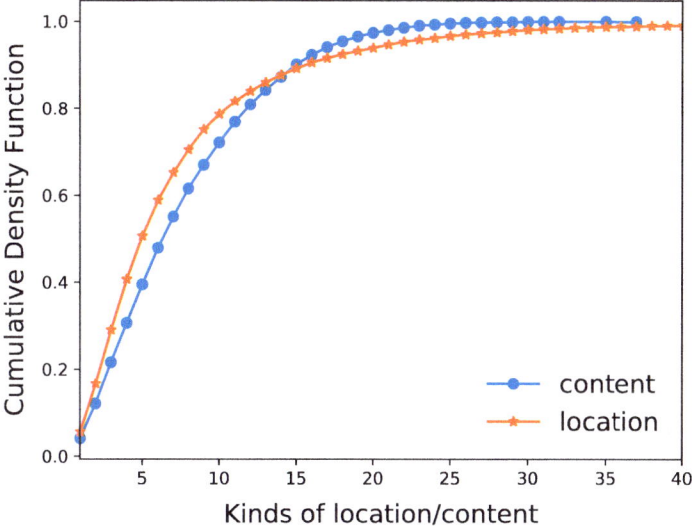

Fig. 5.9   The diversity of users' access behavior in terms of location and content features

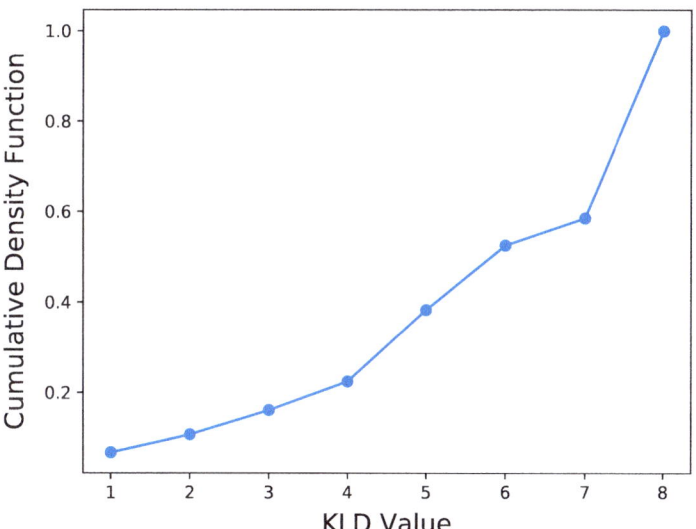

Fig. 5.10   The distribution of the KLD values for 4000 users

As shown in Fig. 5.11, the cumulative density function (CDF) of conditional entropy is relatively lower than that of Shannon entropy for both content and location features. Thus, the predictability of the content feature can be improved by using the location feature as prior knowledge. Likewise, the predictability of the location feature can be improved by using the content feature as prior knowledge.

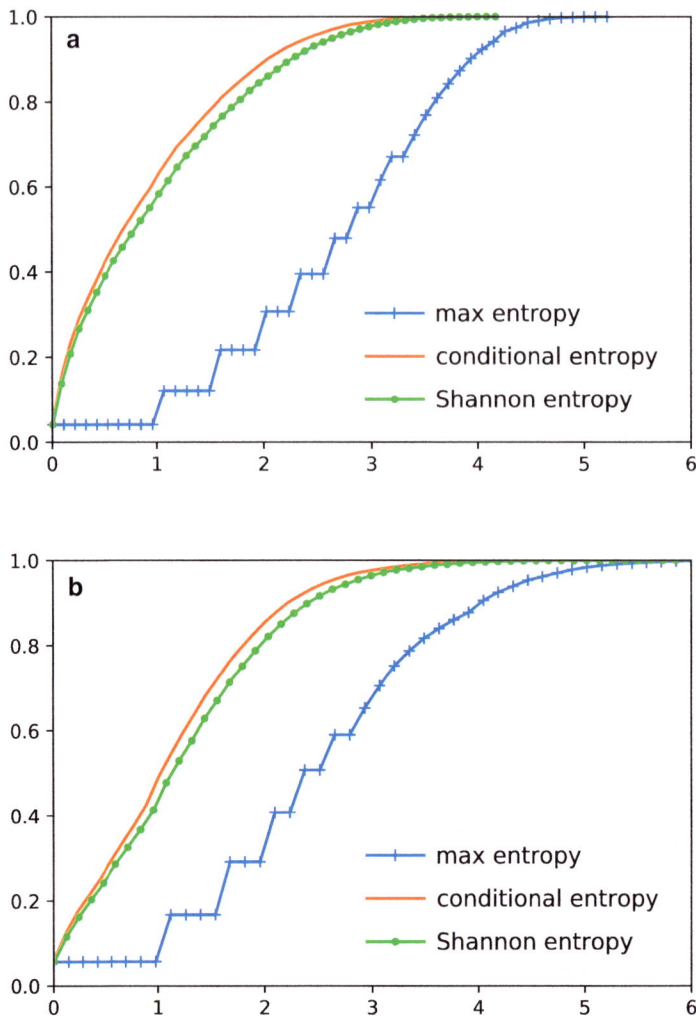

Fig. 5.11 The cumulative density functions of the Shannon entropy, conditional entropy, and maximum entropy of the (**a**) content and (**b**) location features

Next, we are interested in the group behavior of users. Mutual activities and information exchange in social penetration has been discussed in [45], which indicates aggregation among users. For the sake of fairness, we use the normalized entropy (NE) to evaluate the predictability of individual users and user groups.

After confirming the group user behavior, we come to the exact benefit of using user groups instead of individuals. We use the normalized entropy to measure the gap between these two strategies. Figure 5.12 shows the CDF of the normalized entropy of individual users and user groups, Fig. 5.12a shows the normalized

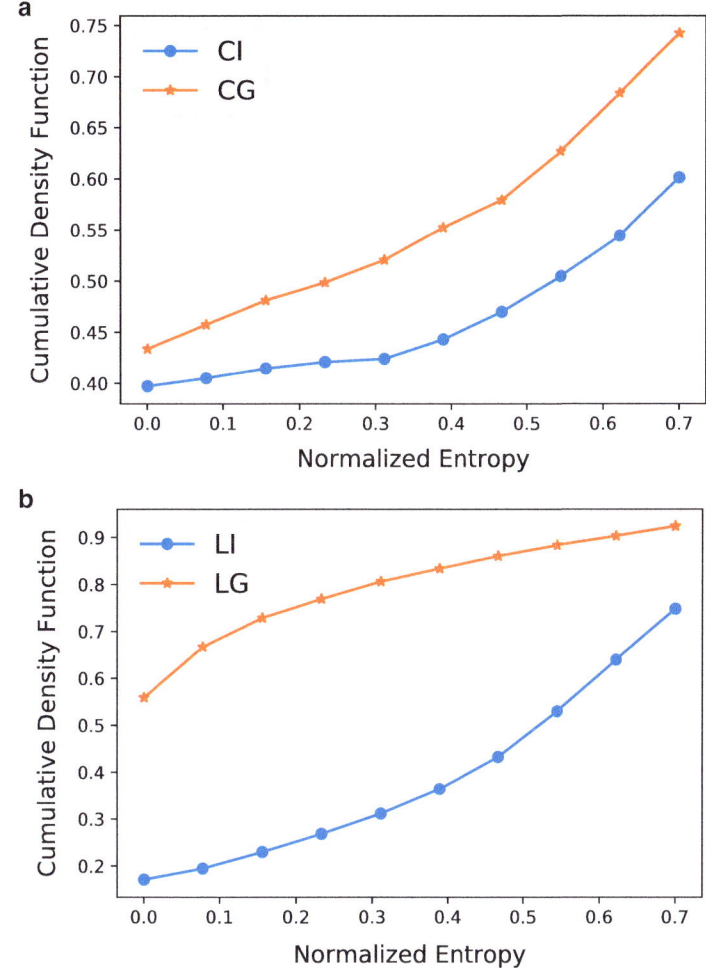

Fig. 5.12 The normalized entropy of the content and location features for user groups and individual users. (**a**) The normalized entropy of the content feature in user groups (CG) and individual users (CI). (**b**) The normalized entropy of the location feature in user groups (LG) and individual users (LI)

entropy of the content feature, and Fig. 5.12b shows the normalized entropy of the location feature. Here, we use the group obtained in Sect. 5.2.4.

According to the above analysis, the locations and contents visited by a user are limited. Thus, good performance can be achieved with limited resources. Users show great diversity of group behavior according to their preferences. Additionally, different features have positive influences on each other during the prediction of user behavior, so a combination of features may result in considerable improvement. Based on these phenomena, we propose a GUCC method in which both the location and content features are considered to form a user similarity matrix, and user groups are divided according to the similarity matrix. Finally, the user groups are used to determine the collaborative relationships among RRHs, and the caching is arranged according to the relationship.

5.2.3   System Model and Problem Formulation

5.2.3.1   System Model

As shown in Fig. 5.13, we discuss the system based on the infrastructure in [46].

In this chapter, we assume that there are M contents in the Internet, denoted as $C = \{C_1, C_2, \ldots, C_M\}$ with sizes $S = \{S_1, S_2, \ldots, S_M\}$. We consider a set of RRHs $B = \{B_1, B_2, \ldots, B_N\}$, where B_i refers to the i-th RRH. Each RRH is equipped with limited local caching capacity, and the size of the local caches is $V = \{V_1, V_2, \ldots, V_N\}$. The cache matrix is formed in Eq. (5.12).

$$\begin{bmatrix} Q_{11} & \cdots & Q_{1M} \\ \vdots & \ddots & \vdots \\ Q_{N1} & \cdots & Q_{NM} \end{bmatrix} \tag{5.12}$$

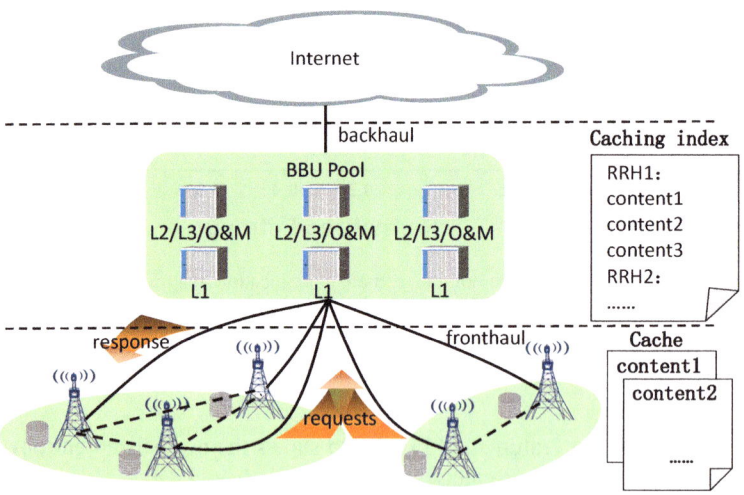

Fig. 5.13 The system infrastructure. The bottom layer of the system is formed by RRHs. Requests are sent to different RRHs based on the users' locations, and the connected RRH is responsible for providing the content to the user from the local cache or by obtaining it elsewhere. The second layer is formed by the building baseband unit (BBU) pool which is the key unit controlling the collaborative relationships among all the RRHs. The third layer is the Internet, which is responsible for the core business

where $Q_{ij} = 1$ indicates the existence of C_j in the local cache of B_i, and $Q_{ij} = 0$ indicates that C_j is not cached on B_i.

In this chapter, we consider a collaborative method, in which T disjoint RRH groups are formed after several procedures, namely $\phi = \{\phi_1, \phi_2, \ldots, \phi_T\}$, and each group consists of a certain number of RRHs. The request matrix is formed in Eq. (5.13).

$$
\begin{bmatrix}
P_{11} & \cdots & P_{1M} \\
\vdots & \ddots & \vdots \\
P_{N1} & \cdots & P_{NM}
\end{bmatrix}
\tag{5.13}
$$

where P_{ij} refers to the probability that users request B_i for C_j.

When a request is received by an RRH (local RRH), the RRH first checks its local cache for the content. If the content exists in its local cache, it is sent directly to the user. Otherwise, the BBU pool receives a request and searches the caching index for a copy. Then, a copy is obtained from the collaborative RRH or the Internet.

5.2.3.2   Problem Formulation

Generally, we are interested in a service with both better quality and lower cost. Therefore, our problem can be divided into two parts, one of which is the hit rate of the requests, and the other is the transmission cost. We define the hit rate as the percentage of requests that can be responded to by the RRHs, whether a local RRH or a collaborative RRH. The hit rate can be formulated as in Eq. (5.14).

$$
\text{Hit}(Q) = \sum_{i=1}^{N} \sum_{j=1}^{M} P_{ij} \cdot Q_{ij} + P_{ij} \cdot (1 - Q_{ij}) \cdot
$$

$$
\left(1 - \prod_{k \in \phi^i, k \neq i} (1 - Q_{kj}) \right)
$$

$$
\text{s.t.} \sum_{j=1}^{M} Q_{ij} S_j \leq V_i, i \in [1, N]
\tag{5.14}
$$

where $\left(1 - \prod_{k \in \phi_i, k \neq i} (1 - Q_{kj}) \right)$ is the indicator function, which means C_j is in the local cache of one of the RRHs in ϕ^i, and ϕ^i is the RRHs that are in the same group as B_i. $P_{ij} \cdot Q_{ij}$ indicates that the request for C_j on B_i can be responded to by B_i, and $P_{ij} \cdot (1 - Q_{ij}) \cdot \left(1 - \prod_{k \in \varphi_i, k \neq i} (1 - Q_{kj}) \right)$ indicates that the request for C_j on B_i is responded to by an RRH in ϕ^i rather than B_i. This constraint represents the cache capacity constraint of RRHs.

We define the transmission cost of the requests in two separate parts, one is the transmission cost among collaborative RRHs, namely Cost_{RRH}, and the other is the transmission cost of obtaining contents from the Internet, namely $\text{Cost}_{\text{Inter}}$. Without loss of generality, we assume that the cost of the transmission between the local cache and users is 0 and we have $\text{Cost}_{\text{RRH}} \leq \text{Cost}_{\text{Inter}}$ for all RRHs, so we set $\text{Cost}_{\text{Inter}} = 1$ and $\text{Cost}_{\text{RRH}} = \gamma$, where $0 < \gamma \leq 1$. Thus, the total transmission cost can be calculated as in (5.15).

$$\text{Cost}\,(Q) = \sum_{i=1}^{N} \sum_{j=1}^{M} P_{ij} \cdot \prod_{k \in \phi^i} \left(1 - Q_{kj}\right) + P_{ij} \cdot \left(1 - Q_{ij}\right)$$

$$\cdot \left(1 - \prod_{k \in \phi^i, k \neq i} \left(1 - Q_{kj}\right)\right) \cdot \gamma \qquad (5.15)$$

where $\prod_{k \in \phi^i} \left(1 - Q_{kj}\right)$ represents the total cost of obtaining content from the Internet and $\left(1 - Q_{ij}\right) \cdot \left(1 - \prod_{k \in \phi^i, k \neq i} \left(1 - Q_{kj}\right)\right) \cdot \gamma$ represents the total cost of obtaining content from collaborative RRHs.

5.2.4 Group User Behavior-Aware Collaborative Caching Scheme

5.2.4.1 Group User Behavior Description

Content and location information are necessary for decision-making. Thus, we choose the content similarity network and the location similarity network to represent the features. The content similarity network and the location similarity network are defined in Eqs. (5.16) and (5.17).

$$S_c = \begin{bmatrix} S_c^{11} & \cdots & S_c^{1N} \\ \vdots & \ddots & \vdots \\ S_c^{N1} & \cdots & S_c^{NN} \end{bmatrix}, \qquad (5.16)$$

$$S_l = \begin{bmatrix} S_l^{11} & \cdots & S_l^{1N} \\ \vdots & \ddots & \vdots \\ S_l^{N1} & \cdots & S_l^{NN} \end{bmatrix} \qquad (5.17)$$

where S_c is the content similarity matrix and S_l is the location similarity matrix. $S_c^{ij}(S_l^{ij})$ is the similarity of user i and user j in terms of the content (location) feature.

As the content and location features are both tagged features, we use Jaccard similarity to form the networks. Let $N(u)$ represent the feature vector of user u and $N(v)$ represent the feature vector of user v. The classic Jaccard similarity is defined in Eq. (5.18).

$$\text{Jaccard}(u, v) = \frac{|N(u) \cap N(v)|}{|N(u) \cup N(v)|} \tag{5.18}$$

Jaccard $(u, v) \in [0, 1]$, and a larger value of Jaccard (u, v) indicates a higher similarity between u and v. However, the classic Jaccard similarity does not consider the popularity of features. Therefore, we change the numerator as shown in Eq. (5.19).

$$\sum_{i \in N(u) \cap N(v)} \frac{1}{\log(1 + N(i))} \tag{5.19}$$

In this chapter, we use a similarity network fusion algorithm for the combination. Traditional similarity network fusion algorithms include SNF [47] and the concatenate algorithm [48]. Despite their tremendous advantages, the concatenate algorithm and SNF may not be suitable in this scenario because the fused similarity network completely relies on the convergence, regardless of the characteristics of different features. Thus, we apply the WSNF algorithm, in which the weight of different features can be adapted according to the application requirements during fusion, to express this distinction.

Essentially, we desire a fused similarity network that can best express all the original similarity networks. As we can regard the network as a hyperspace, the sum of the weighted distances between the target hyperspace and the original hyperspace should be as small as possible. Thus, in WSNF, we solve objective function in Eq. (5.20) to obtain the fused network.

$$\begin{aligned} \min \quad & P(S) = \alpha \|S - S_l\|_2^2 + \beta \|S - S_c\|_2^2 + \lambda S_1 \\ \text{s.t.} \quad & \alpha + \beta = 1, \\ & 0 < \alpha < 1, 0 < \beta < 1, 0 < \lambda < 1 \end{aligned} \tag{5.20}$$

where S_c (S_l) is the original content (location) similarity matrix and S is the fused matrix. Additionally, we use the sum of two 2-norm expressions to calculate the distance described above. Notably, factors α and β are added to scale the weights of different distances to obtain a representation of the importance of the different features.

In this chapter, the above objective function is solved via the fast iterative shrinkage-thresholding algorithm (FISTA) [49].

In this objective function, $\lambda \|S\|_1$ is a continuous convex function, and $\alpha \|S - S_l\|_2^2 + \beta \|S - S_c\|_2^2$ is continuously differentiable with Lipschitz continuous gradient

$L(f)$. Let $f(S) = \alpha||S - S_l||_2^2 + \beta||S - S_c||_2^2$ and $g(S) = \lambda||S||_1$; thus, we have Eq. (5.21).

$$||\nabla f(S') - \nabla f(S)||_2^2 \le L(f)||S' - S||_2^2. \qquad (5.21)$$

$P(S)$ is clearly a convex function. Consider the quadratic approximation of $P(S)$ at a given point S in Eq. (5.22).

$$\hat{P}(S) \cong P(S) + \langle \nabla f(S), S' - S \rangle + \frac{L}{2}||S' - S||^2$$

$$= \frac{L}{2}\left|\left|S' - \left(S - \frac{1}{L}\nabla f(S)\right)\right|\right|_2^2 + \text{const} \qquad (5.22)$$

where L is the Lipschitz constant.

Thus, the iterative function is given in Eq. (5.23) (ignoring constant terms).

$$S_{k+1} = \underset{S}{\text{argmin}} \left\{ \frac{L}{2}\left|\left|S - \left(S_k - \frac{1}{L}\nabla f(S_k)\right)\right|\right|_2^2 + \lambda||S||_1 \right\} \qquad (5.23)$$

Let $B_k = S_k - \frac{1}{L}\nabla f(S_k)$; then, we have Eq. (5.24).

$$S_{k+1} = \underset{x}{\text{argmin}} \left\{ \frac{L}{2}||S - B_k||_2^2 + \lambda||S||_1 \right\} \qquad (5.24)$$

According to the soft-threshold method, we have Eq. (5.25).

$$\frac{L}{2}S - B_k{}_2^2 + \lambda S_1 = \left[\frac{L}{2}(s_1 - b_{k1})^2 + \lambda s_1\right]$$

$$+ \left[\frac{L}{2}(s_2 - b_{k2})^2 + \lambda s_2\right] + \cdots$$

$$+ \left[\frac{L}{2}(s_n - b_{kn})^2 + \lambda s_n\right] \qquad (5.25)$$

Thus, the solution is obtained in Eq. (5.26).

$$S_{k+1} = B_k - \frac{\lambda}{L}\text{sgn}(S_k) \qquad (5.26)$$

A typical iteration of FISTA is shown in Eqs. (5.27)–(5.30).

$$t_{k+1} = \frac{1 + \sqrt{1 + 4t_k^2}}{2}, \tag{5.27}$$

$$y_{k+1} = x_k + \left(\frac{t_k - 1}{t_{k+1}} \right) (x_k - x_{k-1}), \tag{5.28}$$

$$B_{k+1} = y_{k+1} - \frac{1}{L} \nabla f(y_{k+1}), \tag{5.29}$$

$$x_{k+2} = B_{k+1} - \frac{\lambda}{L} \text{sgn}(x_{k+2}) \tag{5.30}$$

The WSNF algorithm is shown in Algorithm 1.

The key advantage of the WSNF algorithm is the adjustable weights, namely α, β, and λ. In this chapter, as the number of users who have visited approximately five locations is 1.5 times as much as that of contents and the gap of entropy between the group and individual in terms of the location feature is 4 times as large as that of the content feature, a reasonable range for α and β may satisfy $\frac{\alpha}{\beta} \in [1.5, 4]$. Therefore, better performance will be achieved when we set $\alpha \in [0.6, 0.8]$.

Algorithm 1 WSNF

---

input: records;fused_metric;win
output: P
1: **function** WSNF(records,fused_metric,win)
2:     $S \leftarrow \emptyset$
3:     **for** metric in fused_metric **do**
4:         Calculate the similarity matrix S_metric of
5:         features in win days metric
6:         nrow,ncol =size(S_metric)
7:         S_metric=reshape(S_metric,nrow × ncol, 1)
8:         S.append(S_metric)
9:     **end for**
10:     Modeling according to (5.27)–(5.30) based on S
11:     $y_1 = x_0$; $t_1 = 1$; $k = 1$; $L = L(f)$
12:     **while** $k \leq$ max_iter **do**
13:         $B_k = y_k - \frac{1}{L} \nabla f(y_k)$
14:         $x_k = B_k - \frac{\lambda}{L} \text{sgn}(y_k)$
15:         $t_{k+1} = \frac{1}{2} \left(1 + \sqrt{1 + 4k^2} \right)$
16:         $y_{k+1} = x_k + \left(\frac{t_k-1}{t_{k+1}} \right) (x_k - x_{k-1})$
17:         $k = k + 1$
18:     **end while**
19:     P=reshape(x_k,nrow,ncol)
20:     **return** P
21: **end function**

5.2.4.2   Cooperative Caching Placement

In this chapter, we will design the collaborative relationship among the RRHs based on user groups rather than individuals. The user groups are determined according to the fused similarity network.

No prior knowledge is available to determine the number of groups, so we use the Louvain community detection algorithm [50] to discover user groups. We denote user groups as $\varphi = \{\varphi_1, \varphi_2, \ldots, \varphi_T\}$. Generally, RRHs that serve the same group should be considered to be collaborative RRHs. However, overlap occurs when an RRH is visited by several groups. To avoid this situation, we adopt the "majority voting" strategy to determine the collaborative relationships among RRHs.

As mentioned before, the caching placement is obtained by solving the optimization problem in Eq. (5.31), where σ is the regulation constant.

$$
\begin{aligned}
&\underset{Q}{\mathrm{argmin}} \quad \sigma\, \mathrm{Cost}\,(Q) - \mathrm{Hit}\,(Q) \\
&\text{s.t.} \qquad \sum_{j=1}^{M} Q_{ij} S_j \leq V_i, \qquad i \in [1, N], \\
&\qquad\qquad Q_{ij} \in \{0, 1\}
\end{aligned}
\tag{5.31}
$$

We transform the problem in Eq. (5.31) to a consistent form in Eq. (5.32).

$$
\begin{aligned}
&\underset{Q}{\mathrm{argmin}} \quad L\,(Q) \\
&\text{s.t.} \qquad Q \in \{0, 1\}
\end{aligned}
\tag{5.32}
$$

where $L\,(Q) = \sigma\, \mathrm{Cost}\,(Q) - \mathrm{Hit}\,(Q) + \lambda F\,(Q)$, where $F\,(Q)$ is the penalty item, $F\,(Q) = \sum_{i=1}^{N}\left\{\max\left[\sum_{j=1}^{M} Q_{ij} S_j - V_i\right]\right\}$, and λ is the penalty factor.

The optimal solution of problem in Eq. (5.32) can be analyzed by MCMC algorithm, the peculiarities of this problem is we need explore whether the use of collaborative relationship between RRHs when cache a certain content can optimize the problem. For a certain content, if the transmission cost of local RRH obtains it from collaborative RRH less than the benefits of using the space saved for caching other content, we say that using the collaborative relationship between RRHs here will optimize the problem.

Before designing the parameters of our MCMC algorithm, some properties of it are worth highlighting. First, the state probability distribution of Markov chain does not need normalization. We only need to know the state distribution up to a constant of proportionality [51]. Second, because each group of RRHs does not affect each other, it is easy to simulate independent Markov chains in parallel, each Markov chain corresponds to a group of RRHs.

The rank of various states $Q^{\{0, 1 \cdots\}}$ in the Markov chain should be related to the value of $L(Q^{\{0, 1 \cdots\}})$, and we will record $L(Q^{\{0, 1 \cdots\}})$ as $L_{\{0, 1 \cdots\}}$ for simplicity in the following sections. We define the rank of Q^s as $v(Q^s) = e^{\beta_1 (L_0 - L_s)}$. Similarly, we define the probability of state Q^s transfer to state Q^t as $q(Q^t \mid Q^s) = \omega e^{\beta_2 (L_s - L_t)}$, where ω is the normalization factor and $\beta_{1,2}$ are the scaling factors. The reason for this is if $L_s > L_t$, we estimate that state Q^t is better than Q^s, and then $v(Q^s) < Q^t$, $q(Q^t \mid Q^s) > q(Q^s \mid Q^t)$.

Algorithm 2 presents the MCMC algorithm in this chapter.

In practical applications, the cache in each RRH should be refreshed periodically. Users are more limited in terms of location features than content features, which means the location feature is more consistent. Thus, the collaborative relationship is relatively stable for a long time. On the basis of this property, we can use a simple refreshing policy, in which only the contents are occasionally refreshed, and the collaborative relationship remains constant for a long time. The cost of the refreshing process includes the transmission cost in the backhaul and the fronthaul, similar to existing caching methods [52].

Algorithm 2 MCMC

---

input: initial state Q^0; transferring function $q(Q^t \mid Q^s)$
output: optimal state Q^*
1: initialize with Q^0
2: **while** sampling times $k <$ pre-defined value K **do**
3:     generate $Q^* \sim q(Q^* \mid Q^k)$
4:     generate $u \sim \text{Uniform}[0, 1]$
5:     **if** $u < \alpha(Q^*, Q^k) = \min\left\{1, \frac{v(Q^*)q(Q^k|Q^*)}{v(Q^k)q(Q^*|Q^k)}\right\}$ **then**
6:         $Q^{k+1} = Q^*$
7:     **else**
8:         $Q^{k+1} = Q^k$
9:     **end if**
10: **end while**
11: **return** Q^*

5.2.5   Numerical Results

5.2.5.1   Description of Dataset

We use the UDR dataset introduced in Sect. 5.2.2 to evaluate the proposed algorithm.

We select the 4000, 6000, and 8000 users who generate the most records as the datasets for the experiments.

We compare WSNF with two reference algorithms, namely the concatenate algorithm and SNF.

We compare GUCC with five reference algorithms, starting with the cache most popular (CMP). Previous studies have proved that CMP outperforms most of the non-collaborative caching methods [33, 35, 52]. To demonstrate the advantages of GUCC from different perspectives, we adopt three algorithms, namely cooperative caching based on location (CCBL), cooperative caching based on content (CCBC), and cooperative caching based on RRH (CCBR). Each algorithm replaces one step in GUCC with a different procedure. CCBL and CCBC use the location similarity network or content similarity network instead of the fused similarity network. CCBR uses the location and content features of the RRHs instead of those of users. The last reference algorithm is ratio collaborative caching (RCC), which combines the non-collaborative caching method in [53] with the collaborative relationship designed by GUCC. In RCC, the contents to be cached are selected randomly, but contents with higher preference have higher probability of being cached.

5.2.5.2 Performance Evaluation of WSNF

We analyze the performance of WSNF based on cross entropy (CE) and modularity (Q).

As the fused similarity matrix is used for community detection, we use modularity and the number of groups to describe the performance.

Let $\beta = 1 - \alpha$ and $\lambda = 0.5$. The performance is evaluated for $\alpha = 0.1, 0.2, \ldots, 0.9$. We compare three algorithms: WSNF, SNF, and concatenate algorithm. The performance of the three algorithms is shown in Fig. 5.14 for different numbers of users (4000, 6000, and 8000).

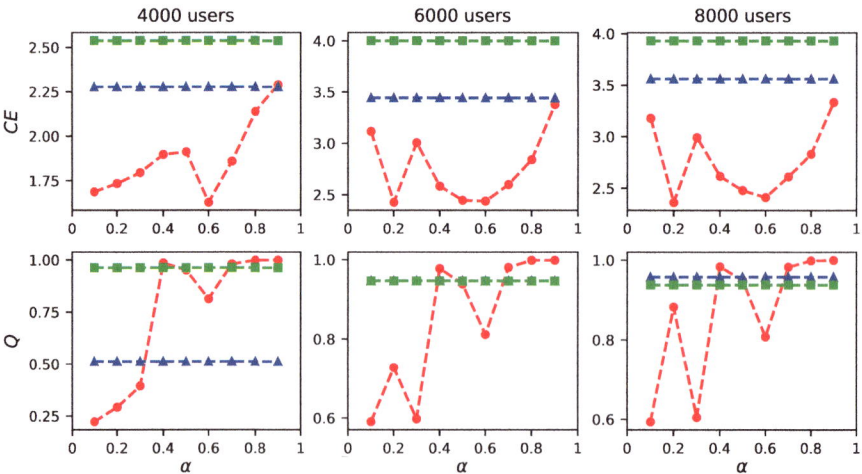

Fig. 5.14 The cross entropy (CE) and modularity (Q) of WSNF, SNF, and concatenate algorithm

In terms of CE, WSNF outperforms the other algorithms. However, as the value of α changes, the CE value varies dramatically. The same variation occurs with the Q values. One possible explanation is that although the value of α greatly affects the performance of WSNF, the effect is linear and should be fitted by a complex model. Taking all the indicators into consideration, WSNF achieves the best performance with $\alpha = 0.6$; therefore, we adopt this setting in the following analysis.

5.2.5.3 Performance Evaluation of GUCC

For simplicity, we assume that each content is the same size and that each RRH has the same cache size. To evaluate our method, we compare the transmission cost and hit rate of GUCC against those of the previously mentioned methods.

Figure 5.15 plots the total hit rate and local hit rate under all six methods for different numbers of users (4000, 6000, and 8000) in the system. The total hit rate is the hit rate defined in Eq. (5.14), and the local hit rate considers only the requests responded to by the local cache.

As shown in the figure, the GUCC has the highest total hit rate and local hit rate, regardless of the number of users. It is worth noting that the performances of CCBL, CCBC, and CCBR are worse than that of GUCC. Thus, the fusion of the similarity network of two features works better than any feature alone, and the features of users provide a better representation of the behavior than those of the RRHs.

To focus on the effect of different methods only, we assume a simple transmission cost expression, where the cost of transmission from the Internet is set to 1, and the cost of transmission from a collaborative RRH is set to γ $(0 < \gamma \leq 1)$. We vary γ in the range of $(0, 1]$ to determine the total transmission cost under different conditions.

The transmission cost is shown in Fig. 5.16. When considering the transmission cost, GUCC still outperforms the other methods. With no collaborative relationship, CMP obtains a stable cost that is not influenced by γ. While the performances of CCBL, CCBC, and CCBR are nearly stable at a high cost with different γ, the curve of the RCC shows a steep gradient.

According to the above results, GUCC achieves performance gains in both hit rate and transmission cost due to the proper design and utilization of the collaborative relationships. By contrast, excessive utilization leads to higher transmission cost, and a lack of utilization results in a lower hit rate. Nevertheless, user behavior can be stable for a period of time, so the caching placement can be effective for a long time.

5.2.6 Conclusion

The service requirements for cellular networks in smart cities have led to increasingly rigorous transmission pressure, and numerous caching methods based on

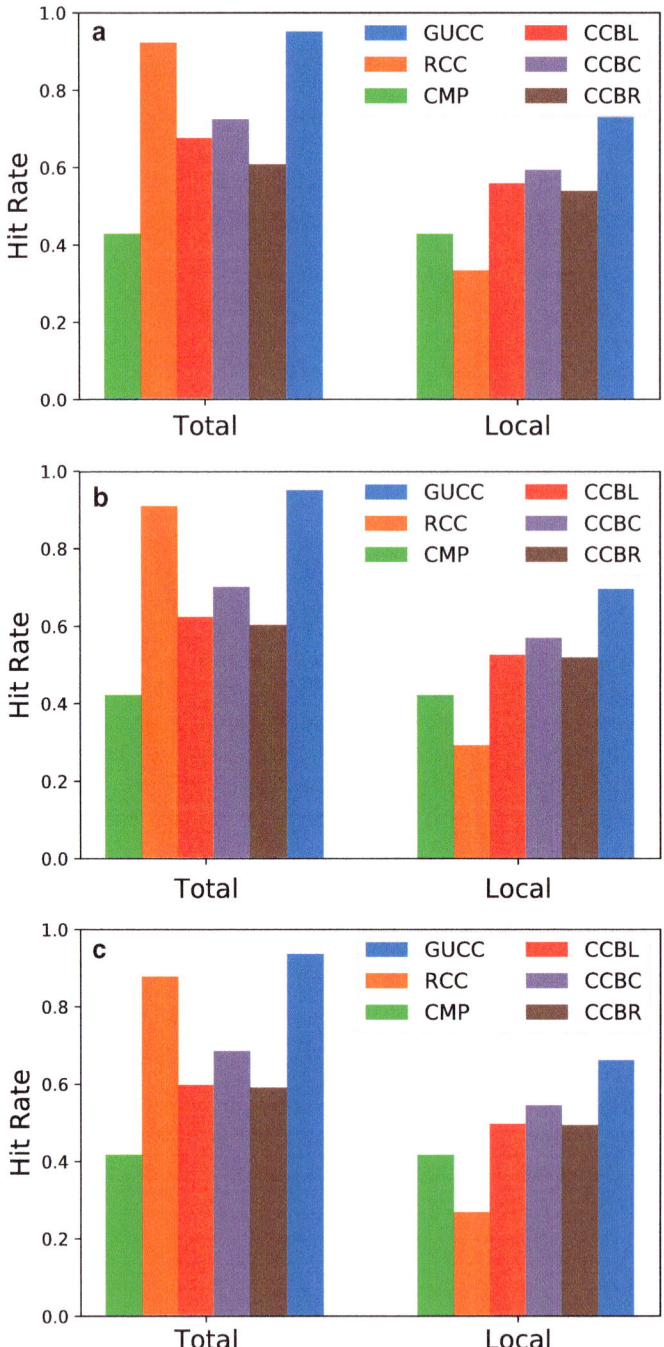

Fig. 5.15 The total hit rates and local hit rates for 4000, 6000, and 8000 users. (**a**) The total hit rates and local hit rates for 4000 users. (**b**) The total hit rates and local hit rates for 6000 users. (**c**) The total hit rates and local hit rates for 8000 users

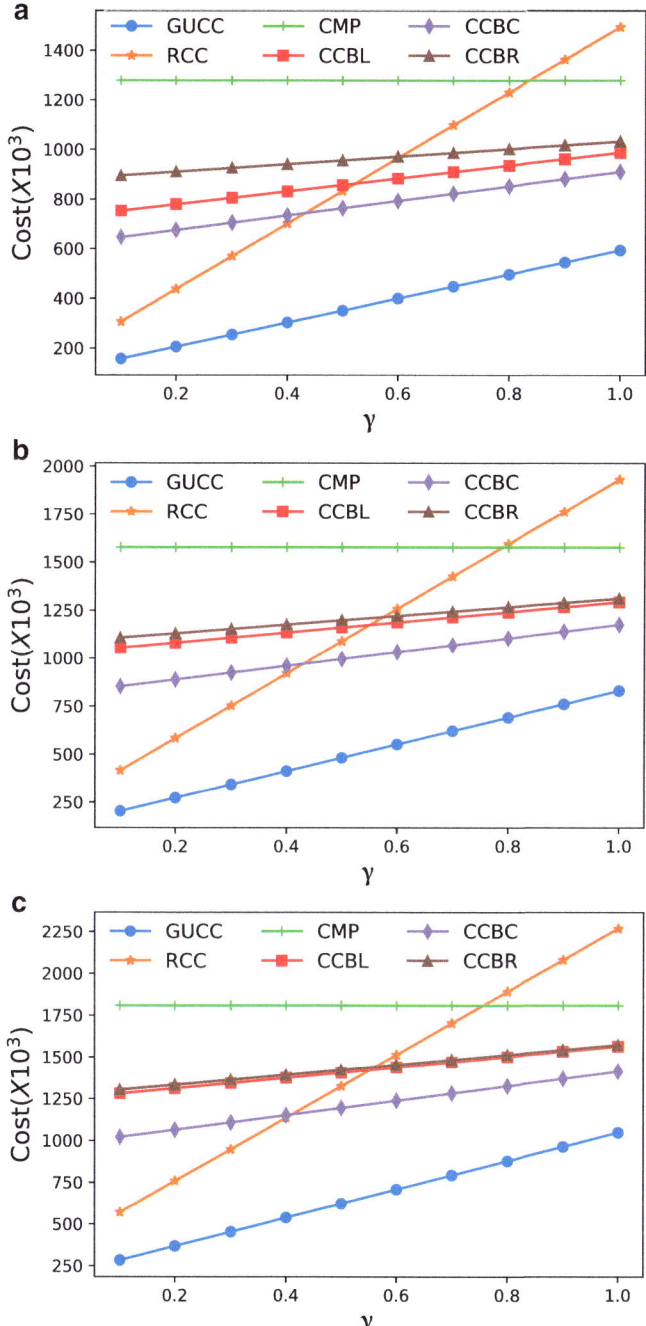

Fig. 5.16 The transmission cost for 4000, 6000, and 8000 users. (**a**) The transmission cost for 4000 users. (**b**) The transmission cost for 6000 users. (**c**) The transmission cost for 8000 users

mobile networks have been proposed to alleviate the pressure. In this chapter, we propose a GUCC method. On the basis of a thorough analysis of UDRs, we find that most users access contents and locations within a limited range. We also find that a combination of user features and a group detection of users can effectively improve the predictability of users. Based on these results, we design a collaborative relationship model. The collaborative relationships among RRHs are formed by "majority voting" in the user groups. During this process, we implement WSNF as an adaptive solution to the large-scale similarity network fusion. After obtaining the collaborative relationship, we propose the caching method by solving an optimization problem, in which both the hit rate and the transmission cost are considered. The problem is analyzed and solved according to MCMC method. The performance gain of the proposed caching method is validated with a large dataset generated from real networks and is compared with several representative caching methods. The results show that our method outperforms the other methods in terms of both hit rate and transmission cost.

References

1. Haught, M.J., Wei, R., Xuerui, Y., Zhang, J.: Understanding the psychology of mobile phone use and mobile shopping of the 1990s cohort in China: a lifestyle approach. In: Mobile Commerce: Concepts, Methodologies, Tools, and Applications, pp. 88–105. IGI Global, Hershey (2018)
2. CNNIC: The 40th China Statistical Report on Internet Development, vol. 7 (2017)
3. Parwez, M.S., Rawat, D.B., Garuba, M.: Big data analytics for user-activity analysis and user-anomaly detection in mobile wireless network. IEEE Trans. Ind. Inf. 13(4), 2058–2065 (2017)
4. Pulselli, R., Ramono, P., Ratti, C., Tiezzi, E.: Computing urban mobile landscapes through monitoring population density based on cellphone chatting. Int. J. Des. Nat. Ecodynamics 3(2), 121–134 (2008)
5. Reades, J., Calabrese, F., Sevtsuk, A., Ratti, C.: Cellular census: explorations in urban data collection. IEEE Pervasive Comput. 6(3), 30–38 (2007)
6. Calabrese, F., Pereira, F.C., Di Lorenzo, G., Liu, L., Ratti, C.: The geography of taste: analyzing cell-phone mobility and social events. In: International conference on Pervasive Computing, pp. 22–37. Springer, Berlin (2010)
7. Girardin, F., Vaccari, A., Gerber, A., Biderman, A., Ratti, C.: Towards estimating the presence of visitors from the aggregate mobile phone network activity they generate. In: International Conference on Computers in Urban Planning and Urban Management (2009)
8. Luo, S., Morone, F., Sarraute, C., Travizano, M., Makse, H.A.: Inferring personal economic status from social network location. Nature Commun. 8, 15227 (2017)
9. Decuyper, A., Rutherford, A., Wadhwa, A., Bauer, J.-M., Krings, G., Gutierrez, T., Blondel, V.D., Luengo-Oroz, M.A.: Estimating food consumption and poverty indices with mobile phone data (2014). arXiv preprint arXiv:1412.2595
10. Park, J., Lee, D.-S., Gonzalez, M.C.: The eigenmode analysis of human motion. J. Stat. Mech: Theory Exp. 2010(11), 11021 (2010)
11. Cole, M.J., Hendahewa, C., Belkin, N.J., Shah, C.: User activity patterns during information search. ACM Trans. Inf. Syst. 33(1), 1 (2015)
12. Zhang, X., Wang, C., Li, Z., Zhu, J., Shi, W., Wang, Q.: Exploring the sequential usage patterns of mobile Internet services based on Markov models. Electron. Commer. Res. Appl. 17, 1–11 (2016)

13. Qiao, Y., Zhao, X., Yang, J., Liu, J.: Mobile big-data-driven rating framework: measuring the relationship between human mobility and app usage behavior. IEEE Netw. **30**(3), 14–21 (2016)
14. Kawazu, H., Toriumi, F., Takano, M., Wada, K., Fukuda, I.: Analytical method of web user behavior using hidden Markov model. In: IEEE International Conference on Big Data (Big Data), 2016, pp. 2518–2524. IEEE, Piscataway (2016)
15. Shafiq, O., Alhajj, R., Rokne, J.G.: On personalizing web search using social network analysis. Inf. Sci. **314**, 55–76 (2015)
16. Jiang, M., Cui, P., Wang, F., Xu, X., Zhu, W., Yang, S.: FEMA: flexible evolutionary multi-faceted analysis for dynamic behavioral pattern discovery. In: Proceedings of the 20th ACM SIGKDD International Conference on Knowledge Discovery and Data Mining, pp. 1186–1195. ACM, New York (2014)
17. Wu, T.L., Li, Y., Zhou, C., Jiang, H., Qian, X.: Statistic analysis of data access behavior in the mobile Internet. In: IEEE/CIC International Conference on Communications in China (ICCC), 2013, pp.89–93. IEEE, Piscataway (2013)
18. Zhou, C., Jiang, H., Chen, Y., Wu, L., Yi, S.: User interest acquisition by adding home and work related contexts on mobile big data analysis. In: IEEE Conference on Computer Communications Workshops (INFOCOM WKSHPS), 2016, pp.201–206. IEEE, Piscataway (2016)
19. Shanahan, J.G., Dai, L.: Large scale distributed data science using apache spark. In: Proceedings of the 21th ACM SIGKDD International Conference on Knowledge Discovery and Data Mining, pp. 2323–2324. ACM, New York (2015)
20. De Lathauwer, L., De Moor, B., Vandewalle, J.: On the best rank-1 and rank-(r 1, r 2,..., rn) approximation of higher-order tensors. SIAM J. Matrix Anal. Appl. **21**(4), 1324–1342 (2000)
21. Shannon, C.E.: A mathematical theory of communication. ACM SIGMOBILE Mobile Comput. Commun. Rev. **5**(1), 3–55 (2001)
22. Yuan, Y., Raubal, M., Liu, Y.: Correlating mobile phone usage and travel behavior-a case study of Harbin, China. Comput. Environ. Urban. Syst. **36**(2), 118–130 (2012)
23. Xu, Y., Shaw, S.-L., Zhao, Z., Yin, L., Lu, F., Chen, J., Fang, Z., Li, Q.: Another tale of two cities: understanding human activity space using actively tracked cellphone location data. Ann. Am. Assoc. Geogr. **106**(2), 489–502 (2016)
24. Wu, L., Leung, H., Jiang, H., Zheng, H., Ma, L.: Incorporating human movement behavior into the analysis of spatially distributed infrastructure. PLoS One **11**(1), e0147216 (2016)
25. Vicente, M.R., Lopez, A.J.: Assessing the regional digital divide across the European union-27. Telecommun. Policy **35**(3), 220–237 (2011)
26. Beijing Municipal Bureau of Statistics: Statistical communiqué for Beijing's national economic and social development, 2016. Beijing Daily, 003(2017)
27. Jinhua Municipal Bureau of Statistics: Statistical communiqué for Jinhua's national economic and social development, 2016. Jinhua Daily, A05(2017)
28. Su, K., Li, J., Fu, H.: Smart city and the applications. In: International Conference on Electronics, Communications and Control, pp. 1028–1031 (2011)
29. Benevolo, C., Dameri, R.P., DAuria, B.: Smart Mobility in Smart City. Springer, Cham (2016)
30. Cisco: Cisco visual networking index: Global mobile data traffic forecast update, 2016–2021 white paper. http://www.cisco.com/c/en/us/solutions/collateral/serviceprovider/visual-networking-index-vni/mobile-white-paper-c11-520862.html
31. Chen, Z., Lee, J., Quek, T.Q., Kountouris, M.: Cooperative caching and transmission design in cluster-centric small cell networks. IEEE Trans. Wirel. Commun. **16**(5), 3401–3415 (2017)
32. Zhou, B., Cui, Y., Tao, M.: Stochastic content-centric multicast scheduling for cache-enabled heterogeneous cellular networks. IEEE Trans. Wirel. Commun. **15**(9), 6284–6297 (2016)
33. Yang, C., Yao, Y., Chen, Z., Xia, B.: Analysis on cache-enabled wireless heterogeneous networks. IEEE Trans. Wirel. Commun. **15**(1), 131–145 (2016)
34. Khreishah, A., Chakareski, J.: Collaborative caching for multicell-coordinated systems. In: IEEE Conference on Computer Communications Workshops (INFOCOM WKSHPS), 2015, pp. 257–262. IEEE, Piscataway (2015)

35. Park, S.-H., Simeone, O., Shitz, S.S.: Joint optimization of cloud and edge processing for fog radio access networks. IEEE Trans. Wirel. Commun. **15**(11), 7621–7632 (2016)
36. Zhou, B., Cui, Y., Tao, M.: Optimal dynamic multicast scheduling for cache-enabled content-centric wireless networks. IEEE Trans. Commun. **65**(7), 2956–2970 (2017)
37. Tamoor-Ul-Hassan, S., Bennis, M., Nardelli, P.H.J., Latva-Aho, M.: Modeling and analysis of content caching in wireless small cell networks. In: 2015 International Symposium on Wireless Communication Systems (ISWCS), vol. 57(1), pp. 56–60 (2015)
38. Nagaraja, B.B., Nagananda, K.G.: Caching with unknown popularity profiles in small cell networks. In: Global Communications Conference (GLOBECOM), 2015, pp. 1–6. IEEE, Piscataway (2015)
39. Peng, X., Shen, J.-C., Zhang, J., Letaief, K.B.: Backhaul-aware caching placement for wireless networks. In: Global Communications Conference (GLOBECOM), 2015, pp. 1–6. IEEE, Piscataway (2015)
40. Van Mieghem, P., Blenn, N., Doerr, C.: Lognormal distribution in the digg online social network. Eur. Phys. J. B **83**(2), 251–261 (2011)
41. Mahanti, A., Carlsson, N., Mahanti, A., Arlitt, M., Williamson, C.: A tale of the tails: power-laws in Internet measurements. IEEE Netw. **27**(1), 59–64 (2013)
42. Zhang, Y., Yang, L., Jiang, H., Yi, S., Hu, Z.: Mining mobile Internet lifestyles in distinct urban areas: tales of two cities. IEEE Access **6**, 36208–36217 (2018)
43. Zhou, C., Jiang, H., Chen, Y., Wu, J., Zhou, J., Wu, Y.: TCB: a feature transformation method based central behavior for user interest prediction on mobile big data. Int. J. Distrib. Sens. Netw. **12**(10) (2016). https://doi.org/10.1177/1550147716671256
44. Kullback, S.: The Kullback-Leibler distance. Am. Stat. **41**(4), 340–341 (1987)
45. Taylor, D.A.: The development of interpersonal relationships: social penetration processes. J. Soc. Psychol. **75**(1), 79–90 (1968)
46. Peng, M., Yan, S., Zhang, K., Wang, C.: Fog-computing-based radio access networks: issues and challenges. IEEE Netw. **30**(4), 46–53 (2016)
47. Wang, B., Mezlini, A.M., Demir, F., Fiume, M., Tu, Z., Brudno, M., Haibe-Kains, B., Goldenberg, A.: Similarity network fusion for aggregating data types on a genomic scale. Nat. Methods **11**(3), 333–337 (2014)
48. Wang, Z., Zhang, D., Zhou, X., Yang, D., Yu, Z., Yu, Z.: Discovering and profiling overlapping communities in location-based social networks. IEEE Trans. Syst. Man Cybern. Syst. Hum. **44**(4), 499–509 (2014)
49. Beck, A., Teboulle, M.: A fast iterative shrinkage-thresholding algorithm for linear inverse problems. SIAM J. Imag. Sci. **2**(1), 183–202 (2009)
50. Blondel, V.D., Guillaume, J.-L., Lambiotte, R., Lefebvre, E.: Fast unfolding of communities in large networks. J. Stat. Mech. Theory Exp. **2008**(10), 10008 (2008)
51. Andrieu, C., de Freitas, N., Doucet, A., Jordan, M.I.: An introduction to MCMC for machine learning. Mach. Learn. **50**(1), 5–43 (2003)
52. Liu, D., Yang, C.: Energy efficiency of downlink networks with caching at base stations. IEEE J. Sel. Areas Commun. **34**(4), 907–922 (2016)
53. Blaszczyszyn, B., Giovanidis, A.: Optimal geographic caching in cellular networks. In: IEEE International Conference on Communications (ICC), 2015, pp. 3358–3363. IEEE, Piscataway (2015)

Chapter 6
Conclusion, Remarks, and Future Directions

Abstract This chapter is the summary part of the whole book. It mainly sums up the level of development and methods as well as applications of mobile data, mobile data analysis, and mobile data mining. Besides, we introduce research direction of mobile data analysis and mining based on artificial intelligence (AI), Internet of things (IoT), and network architecture.

Keywords Mobile data · Mobile data analysis · Mobile data mining · Artificial intelligence (AI) · Internet of Things (IoT) · Network architecture

Mobile data has become a necessary element of transport in recent years. The integration of mobile communication technology and information technology has made mobile technology the focus of attention in the industry. With the convergence of mobile communication and mobile computing technology, mobile technology has gradually matured. Also, mobile data has been widely used in mobile technology. Mobile data has many applications, such as project estimating demand from mobile network data (EDMOND) has been a major exercise, involving many new techniques to deal with the challenges of data development for such a complex city [1]. The Strategic Analysis Department at Transport for London is responsible for the transport models that are used to forecast future traffic and public transport demand, congestion, and crowding in London. It also considers demand for cycling, walking, taxi, and other travel. In Project EDMOND, TfL worked with Jacobs, AECOM, and Telefonica to develop new data that will be used to help maintain and update TfL's transport planning models and to provide new policy insight using mobile network data from Telefonica's O2 (some 30% of UK mobile users) fused with various other datasets relating to transport in London. Mobile network data used were aggregated to protect the privacy of the users. What's more, mobile data technique can also be used to build systems to protect the environment. An energy-efficient intelligent system based on mobile data technique is built for the treatment of thermal power plants [2].

Big data refers to a collection of data that cannot be captured, managed, and processed by conventional software tools within a certain time frame. The level

© Springer Nature Switzerland AG 2019

H. Jiang et al., *Mobile Data Mining and Applications*, Information Fusion and Data Science, https://doi.org/10.1007/978-3-030-16503-1_6

of data analysis ability determines the quality and success of the value discovery process in big data. Although the concept of big data has been mentioned by more and more media and industry in the past 2 years, the development of big data analysis is still in primary stage. From the perspective of industry practice, only a few enterprises of a few industries can conduct basic analysis and application of big data. From the perspective of technology development, some mature data analysis technologies, such as business intelligence technology and data mining, have been widely applied in many industry fields. The methods of big data analysis include analytic visualizations, data mining algorithms, predictive analytic capabilities, semantic engine, data quality and master data management, data storage, and so on. The application of big data analysis is very extensive, such as financial transaction, improve quality of city, improve law enforcement efficiency, optimize machine and equipment performance, improve medical standards, optimize business processes, and satisfy request of customer. Also, big data analysis is used for physical Internet-based intelligent manufacturing shop floors [3].

Data mining is a step in knowledge discovery in databases (KDD). Data mining generally refers to the process of searching for information hidden in it from a large amount of data through an algorithm. Data mining is often associated with computer science and achieves these goals through statistics, online analysis processing, intelligence retrieval, machine learning, expert systems (reliant on past rules of thumb), and pattern recognition. At present, data mining is been widely used in database for analyzing products' value and demand. The main methods include classification, estimation, prediction, clustering, description, and visualization. Incident reporting and investigation are components of safety management systems. Timely and accurate identification of risk factors is crucial to effective prevention strategies. However, risk factor identification is often hampered by size, complexity, and the need for human involvement in categorizing incident data. An approach based on data mining is designed to identify the risk factors in safety management [4]. Also, drug use motives are relevant to understand substance use among students. Data mining techniques present some advantages that can help to improve our understanding of drug use issue [5].

Mobile big data analysis and data mining have broad application prospects. Next we introduce study direction from three aspects.

6.1 Artificial Intelligence (AI)

Under situations of large volumes of data, artificial intelligence (AI) allows delegation of difficult pattern recognition, learning, and other tasks to computer-based approaches [6]. In addition, AI contributes to the velocity and variety of data due to its high speed data processing ability and powerful comprehension. For mobile big data (MBD), to the best of our knowledge, the current contribution of AI is somewhat incremental, while the application prospect of AI keeps rather promising. Dean et al. [7] and Zhang et al. [8] develop the structure and architecture of large distributed deep networks, which can be trained and evaluated in parallel over a mass of computing devices and would be applicable for MBD. He et al. [9] introduce a unified data model based on the random matrix theory and machine learning to the big data analytics in the mobile cellular networks. Alsheikh et al. [10] propose a Spark-based framework for learning deep models for time-efficient MBD analytics within large-scale mobile systems. According to [10], deep learning, on the one hand, shows several advantages in MBD analytics, including the high accuracy, the ability to generate intrinsic features, etc. On the other hand, the curse of dimensionality, the large demand of data for deep learning networks, and the volatility characteristic of MBD are challenges yet to be tackled.

Analysis and management of data is the main method to improve the effectiveness, but with the increasing number and type of data, the general data processing software cannot meet the characteristics of the current data. The dataset mentioned above is big data, which obviously surpasses the traditional form of MB and GB for storing data. In order to give full play to the role of data guarantee in decision-making process and establish a more scientific decision-making system, it is necessary to continuously optimize the data processing mode in the process of calculating data. Big data has abundant data types and contents. It is transmitted in microseconds and milliseconds, reflecting its fast characteristics. In addition to that, the operation process of classifying large data is difficult, so it is necessary to effectively eliminate the influencing factors to ensure the access to real and reliable information.

Machine learning is a common method which includes many methods. The first one is regression algorithm. Regression algorithm is an important prerequisite for generating other algorithms which includes linear and logical regression. The former mainly solves numerical problems, while the latter mainly predicts effective numbers. This method is also a classification algorithm, so it often obtains the results of discrete classification. The second one is neural network. The main principle of neural network is to study and simulate the working process of the brain in an all-round way, forming several hierarchical structures through multiple processing units, and forming a logical framework together and its operation principle is that the receiving layer acquires signals, the hidden layer decomposes and processes data, and the output layer is the final result of reasonable integration.

In addition, every processing unit existing in the network can be considered as a model. If thousands of neural networks exist in the hidden layer, they are called deep neural networks. The research on this basis is deep learning. The third one is support vector machine. Compared with the biological background formed by neural network, support vector machine has a very strong mathematical component. Although the method itself is not only an enhanced logistic regression algorithm, it can effectively merge with Gauss function, successfully map low-dimensional space from high-dimensional space, and smoothly transform linear classification technology. The fourth one is clustering algorithm. Clustering algorithm calculates the distance of population formation scientifically and divides it into several ethnic groups according to distance and proximity. The fifth one is dimension reduction algorithm. Its core feature is to reduce the high dimension to the low dimension.

The research proves that this method can fully save the data, effectively compress the data, improve the operation level of the algorithm, and visualize the processing of relevant data. The sixth one is recommendation algorithm. It is one of the most popular algorithms at present. It is widely used in e-commerce, and can be recommended to people's favorite things according to the purchasing situation.

There are many algorithms in machine learning, among which support vector machines and neural networks are the most important ones. Combining the learning content, the algorithm can be divided into three types. First, all-round supervised learning refers to the computer that obtains information knowledge through the environment, prompts wrong judgments, and provides correct answers. The main purpose of supervised learning is to use computers to accumulate experience and improve skills in the whole learning process, and to correctly solve the problems that have not been studied, such as regression, neural network, and support vector machine. Second, it will not supervise learning reasonably, that is, the computer collects, collates, and screens valuable information on its own in the network, but lacks accurate goals, mainly including clustering and dimensionality reduction. Continuously strengthen learning, without any environmental instructions, the computer can correctly evaluate the predicted answers. Data mining can analyze data more comprehensively and summarize the law of development scientifically. The key steps are preparing data, searching law, and revealing law. In the process of social development, data mining is the most critical content, widely used in electric power, medicine, and agriculture. At present, the continuous strengthening of data mining needs to apply machine learning scientifically based on big data, constantly improve operational capability, and promote sustainable development of data mining. The content of the traditional machine learning algorithms is the key, but in the process of data storage, TB and PB level data cannot be stored in the computer successfully. In the era of big data processing, most algorithms cannot meet the operational requirements of data mining. Therefore, only through continuous improvement can they highly accord with the requirements of current development. In the machine learning algorithms, the artificial neural network method can establish different models to embody the robustness and description function, and strictly control the accuracy. In the era of big data, on the one hand, machine learning becomes more difficult because of the continuous development of the data scale. On the other hand, due to the diversified data distribution, the traditional machine learning requires that data outlier to be distributed independently, which makes it unable to fully play its role. Therefore, we should actively optimize the algorithm to better carry out data mining operations. Machine learning is applied to mobile big data processing, which improves the classification performance. After designing classifiers in sample-intensive areas, it can play an effective role. In the era of big data, machine learning has successfully broken the concept of conceptual learning. Keeping pace with the development of the times has become the key channel to mine and improve data.

6.2   Internet of Things (IoT)

The proliferation of these devices in a communicating-actuating network creates the Internet of things (IoT). The IoT will provide the tools to establish a major global data-driven ecosystem with its emphasis on big data. The massive data generated by the Internet of things (IoT) are considered of high business value, and data mining algorithms can be applied to IoT for extracting hidden information from data. Ericsson recently released the latest mobile market report covering global mobile user trends and development prospects, the development trend of VoLTE as well as that of mobile data traffic and the IoT, global network coverage, and network evolution trends. According to Ericsson's forecast, the number of connected devices worldwide is expected to exceed 30 billion by 2023, of which about 20 billion are IoT devices. Networked IoT terminals include networked cars, machines, meters, sensors, POS machines, consumer electronics, and wearable devices.

With the development of terminal technology and the expansion of the application areas, the application value and service demand of the Internet of things are constantly improving. Efficient mobile data processing technology is considered as a reliable guarantee for the development and application of the IoT. The rationality of energy utilization in the resource-constrained IoT, however, directly affects the reliability of data aggregation and reflects the level of energy efficiency. Therefore, energy efficiency optimization is of great significance to data aggregation of the resource-constrained IoT. However, the massive data in the IoT come from the vast and changeable natural environment and complex human activity areas. In addition, the efficiency of data aggregation and further analysis and processing is inevitably limited by the environment, and even suffers from the negative impact of human activities. Hence, it is very important to introduce human-centered mobile data mining technology into the IoT. To be specific, people acquire data related to cities from mobile intelligent devices and vehicles in daily life. Then, data integration, analysis, and mining are done via data processing technology, after which the knowledge is acquired. Based on the above work, monitoring of air pollution, water quality, and noise, and the planning for road traffic to relieve congestion in urban areas can be effectively realized.

6.3   Network Architecture

With the arrival of 5G, building network architecture based on big data is playing a good role in ensuring the stability of communication, which can make data transmission more reliable and effective. Big data is a key technology of 5G wireless communication network.

Wireless mobile networks are now experiencing explosive increases in data traffic. According to the white paper from Cisco with a global traffic forecast, global mobile traffic increased 63% in 2016 and reached 7.2 EB per month. By 2021,

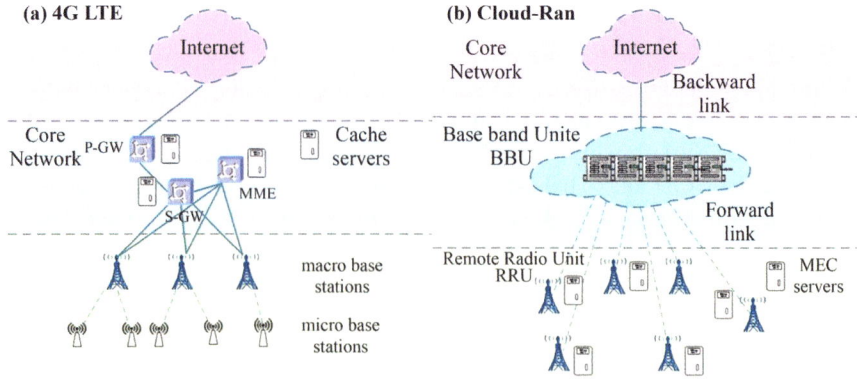

Fig. 6.1 Comparison of 4G and 5G network architectures. (**a**) Left: The traditional 4G network architecture, (**b**) right: Cloud-Ran combining MEC

WiFi and mobile devices will account for 63% of IP traffic. Mobile edge computing (MEC) was proposed by the European Telecommunications Standards Institute (ETSI) in 2014, and it has been viewed as effective solution to alleviate backhaul bandwidth consumption. MEC severs are deployed at the network edges (e.g., macro base stations (BSs) and macro BSs in LTE), which allows sharing and distributing content locally and thus reducing the backhaul overload. Many new 5G architecture, such as Cloud-Ran, Fog-Ran, and content centric networking (CCN), will be built based on mobile big data in order to ensure the stability of communication, which can make data transmission more reliable and effective. Big data is a key technology of 5G wireless communication network. The network architecture is illustrated in Fig. 6.1. As the pictures show, trends of computing architecture begin to change from centralized cloud computing to mobile edge computing (MEC). Mobile edge computing is characterized by pushing services such as mobile computing, network control, and storage from the cloud to the edge of the network, which makes it possible for resource-enhancing and mission-critical applications of mobile devices. The mobile edge computing architecture effectively fills the gap between the IoT device and the back-end computing architecture. The storage and analysis of mobile data can be performed by storing and calculating data at the edge of the network.

References

1. Abdreshova, S., Golubeva, T., Konshin, S., Bahtaev, S., Birlesbek, A.: A mobile data transfer in energy-efficient intelligent systems for wastewater treatment of thermal power plants. In: 2018 6th International Conference on Future Internet of Things and Cloud Workshops (FiCloudW), pp. 164–169 (2018)
2. Loshin, D.: Big Data Analytics: From Strategic Planning to Enterprise Integration with Tools, Techniques, NoSQL, and Graph. Morgan Kaufmann Publishers Inc., San Francisco (2003)

3. Neffendorf, H., Wroe, C., Davies, A.: Enhancing mobile data with understanding of bias and error. Int. J. Mark. Res. **60**(6), 635–639 (2018)
4. Shi, D., Guan, J., Zurada, J., Manikas, A.: A data-mining approach to identification of risk factors in safety management systems. J. Manag. Inf. Syst. **34**(4), 1054–1081 (2017)
5. Jimenez, R., Anupol, J., Cajal, B., Gervilla, E.: Data mining techniques for drug use research. Addict. Behav. Rep. **8**, 128–135 (2018)
6. O'Leary, D.E.: Artificial intelligence and big data. IEEE Intell. Syst. **28**(2), 96–99 (2013)
7. Dean, J., et al.: Large scale distributed deep networks. In: Pereira, F., Burges, C.J.C., Bottou, L., Weinberger, K.Q. (eds.) Advances in Neural Information Processing Systems, vol. 25, pp. 1223–1231. Curran Associates, New York (2012)
8. Zhang, K., Chen, X.-W.: Large-scale deep belief nets with MapReduce. IEEE Access. **2**, 395–403 (2014)
9. He, Y., et al.: Big data analytics in mobile cellular networks. IEEE Access. **4**, 1985–1996 (2016)
10. Alsheikh, M.A., Niyato, D., Lin, S., Tan, H.-p., Han, Z.: Mobile big data analytics using deep learning and apache spark. IEEE Netw. **30**(3), 22–29 (2016)

Index

A
Access points (APs), 54, 56, 60, 66–68, 70–71, 77–81, 93
APP data, 158–164, 166–171, 184, 185, 194
Artificial intelligence (AI), 216–218

B
Base stations (BSs)
    cellular, 121
    communication unit, 122
    D2D server, 100
    DMUES, 121
    energy-saving strategy, 121, 130, 131, 135
    femtocells, 54, 99, 100
    frequency sequences, 105
    hibernation/cell zooming, 122
    infrastructure, 124
    LTE, 71
    network community structure, 133
    SBSs, 54, 75
    spatial distribution, 140
    strength of tie, 31, 45
    UDR, 137
    WiFi networks, 53
Behavior-aware collaborative caching
    group-based user behavior analysis, 196–200
    GUCC method, 195
    problem formulation, 201–202
    RRHs, 195, 196
    system model, 200–201
Behavior discovery
    evaluation
        DT and RF, 38

Euclidean distance, 37
    feature sets, 38
    HS and HT, 41
    $HS/HS_{STA}/HT/HT_{STA}$, 41–43
    performance, 39, 40
    prediction performance comparison, 41
    UDRs, 36–37
    feature transformation, 31–36
    mobile internet usage characteristics, 29–31
    in participatory sensing, 6–8
Brand cluster, 162, 163, 165–172

C
Call detail records (CDR), 140, 181
Cellular data traffic, 53, 54, 75
Collective mobility discovery method based on community differences (CMDCD), 137–138, 143, 144, 152–157
Context-awareness
    evaluation, 49–50
    UIAHW design, 43–46
    user interest, 46–49
Coolpad, 160, 166–168, 171

D
Data mining, mobile
    algorithm, 218
    business intelligence technology, 216
    communication industry, 2
    cross-disciplinary technology, 2
    KDD, 216
    mobile data analysis, 2, 3, 181
    operational requirements, 218
    user loyalty, 3–4

© Springer Nature Switzerland AG 2019
H. Jiang et al., *Mobile Data Mining and Applications*, Information Fusion and Data
Science, https://doi.org/10.1007/978-3-030-16503-1

Printed by Printforce, the Netherlands